Springer Geography

The Springer Geography series seeks to publish a broad portfolio of scientific books, aiming at researchers, students, and everyone interested in geographical research.

The series includes peer-reviewed monographs, edited volumes, textbooks, and conference proceedings. It covers the major topics in geography and geographical sciences including, but not limited to; Economic Geography, Landscape and Urban Planning, Urban Geography, Physical Geography and Environmental Geography.

Springer Geography — now indexed in Scopus

Pintu Mandal · Suman Paul

Climatic Vulnerability and Migration: Evidence from Indian Sundarbans

Springer

Pintu Mandal
Department of Geography
Sidho-Kanho-Birsha University
Purulia, West Bengal, India

Suman Paul
Department of Geography
Sidho-Kanho-Birsha University
Purulia, West Bengal, India

ISSN 2194-315X ISSN 2194-3168 (electronic)
Springer Geography
ISBN 978-3-032-05857-7 ISBN 978-3-032-05858-4 (eBook)
https://doi.org/10.1007/978-3-032-05858-4

This Springer imprint is published by the registered company Springer Nature Switzerland AG
The registered company address is: Gewerbestrasse 11, 6330 Cham, Switzerland

Dedicated to
Young Scholars in the Atmospheric Science,
Sustainability Science, Ecology,
Environmental Science, Hazard and Disaster
Management, Geospatial Science,
Geography and Policy-Makers

Preface

Climate change, environmental vulnerability, and socio-economic fragility have become increasingly intertwined in the Indian Sundarbans, one of the biggest and most environmentally vulnerable deltaic ecosystems in the world. The misery experienced by the coastal people, whose lives and livelihoods are frequent disruptions, increases in tandem with the frequency and intensity of cyclonic storms. The complicated socio-environmental result of these climate-induced vulnerabilities is migration, both temporary and permanent, which has arisen not just as a coping strategy. Given this, the book *Climatic Vulnerability and Migration: Evidence from Indian Sundarbans* offers a methodologically sound and empirically supported investigation of the complex interrelationship between cyclones and migration patterns in this ecologically vulnerable area.

A team of academics dedicated to examining the many facets of vulnerability in climate-sensitive regions has produced this book via years of fieldwork and scholarly investigation. The research uses a combination of primary and secondary data, statistical models, geospatial methods, and field observations to tell a compelling story about how communities in the Indian Sundarbans migrate, adapt, and cope with the destructive effects of frequent cyclonic events. The theoretical and methodological foundation of the book is laid forth in Chap. 1. The main ideas of climate-induced migration and vulnerability are introduced, the study's history and significance are outlined, important research gaps are identified, and the precise goals are outlined. The reader is given a clear understanding of the research framework and its breadth through the clear articulation of the subject field, data sources, and methodological decisions. A thorough bibliometric analysis of the national and international intellectual discourse on climate-driven migration is conducted in Chap. 2. This chapter uses quantitative analysis and visualization approaches to decipher the patterns, collaboration networks, and thematic progression of the current literature. The conclusions drawn from this show not just the subject's increasing significance but also the conceptual gaps that this book aims to fill. The narrative shifts from literature to landscape in Chap. 3. A thorough indicator-based index methodology is used here to quantify the coastal population's vulnerability. The chapter explores how agricultural instability, geographical inequality, and infrastructure deficiencies all increase the likelihood of human migration. Additionally, it provides a comprehensive knowledge of how vulnerability and migratory outcomes are mediated by adaptive capacities—or the absence of them. The living experiences of rural migrants are emphasized in Chap. 4. The migratory patterns, socio-economic traits, and demographic trends of the Sundarbans' coastal residents are captured by means of carefully collected primary data. This chapter explains how cyclonic stress serves as a catalyst that forces people and families to relocate in pursuit of safety, stability, and long-term employment. Chapter 5 uses strong statistical modeling to further explore this question. By determining and measuring the factors that contribute to migration caused by cyclones, it highlights the interaction between environmental exposure, livelihood difficulties, social capital, and demographic factors. It illustrates how different types of capital—natural, social, and economic—influence migration choices, emphasizing that structurally ingrained vulnerabilities rather than chance or whimsy drive movement. Actionable ideas are distilled from the empirical results in Chap. 6. Key findings that are in line with the goals of the study are summarized, and specific policy recommendations are made to lessen migration brought on by cyclones. This chapter describes a multi-pronged policy strategy that includes climate migration into official adaption efforts, improving infrastructure resilience, and creating targeted livelihood interventions. Along with discussing the research's wider scholarly contributions in terms of knowledge generation, methodological innovation, and theoretical progress, it also outlines potential avenues for further study.

At the intersection of climate change, migration studies, disaster management, and regional planning, this book is meant for academics, policymakers, development professionals, and students. By bringing the conversation on climate-induced migration to the Indian Sundarbans, it seeks to close a crucial gap in both scholarship and policy. The present book is fundamentally a plea to rethink climate resilience, not just by improving infrastructure but also by having a comprehensive view of human mobility as a tactical adaptation measure. The Sundarbans' local residents, whose tales of tenacity, hardship, and hope are at the core of our work, deserve to be thanked. Every level of analysis has been enhanced and informed by their voices and life experiences. The academic institutions, data sources, and funding organizations that helped with different stages of this study project are also acknowledged. We really hope this book makes a significant contribution to the scholarly conversation and shapes policymaking in climate-vulnerable areas, not only in India but throughout the world.

Purulia, West Bengal, India

Pintu Mandal
Suman Paul

Acknowledgments We would like to express our deep appreciation to all the Faculty Members at RAC-Department of Geography, Sidho-Kanho-Birsha University for their invaluable support, encouragement, and the motivation they gave us to do this project. We sincerely acknowledge Dr. Subhasis Bhattacharya, Professor, Department of Economics, Sidho-Kanho-Birsha University, Purulia, for his valuable suggestions and the mental support he provided for our research work. We expressed our great respect to Rita Samanta, a member of Zilla Parishad, Pathar Pratima Panchayat Samity, and Dr. Subham Dey, Assistant Director of Agriculture, WBAS Admin, Pathar Pratima CD Block, for providing data, valuable information, and support. We also thank the Director, RMC-IMD—India Meteorological Department, Alipore, and Kolkata, for the climate data. We must express our sincere thanks to the villagers of my study area, without their help, this part of the Ph.D. thesis of Pintu Mandal would not have taken its present shape. Their role in shaping this research is deeply appreciated. We would like to thank our parents, who have always supported us throughout the project.

Purulia, West Bengal, India

Pintu Mandal
Suman Paul

Competing Interests The authors have no competing interests to declare that are relevant to the content of this manuscript.

Contents

About the Authors

Dr. Pintu Mandal was born on April 10, 1994, in Bankura, West Bengal, India, 722137. He graduated (B.A.) with a degree in Geography in 2014 from Sidho-Kanho-Birsha University, West Bengal, India with stand as college topper and completed his Master's (M.A.) in Geography in 2016 from the same university. Currently, he has completed his Ph.D. degree from Sidho-Kanho-Birsha University, West Bengal, India specializing in Environmental Hazards and Disasters Analysis, Climatic Migration, and Remote Sensing. His Ph.D. research is titled "*Cyclone-Induced Vulnerability and Migration: Evidence from Pathar Pratima Block, South 24 Parganas, and West Bengal, India.*" Pintu has presented his research at various international-, national-, and state-level seminars and conferences. He has also published several articles in reputable Scopus-cited journals and edited books.

Dr. Suman Paul was born in Kolkata, West Bengal, India on July 31, 1978. He received his B.Sc. and M.Sc. degrees from University of Calcutta, Calcutta, India in 2000 and 2002 respectively. He was associated as an Assistant Professor at Post-Graduate Department of Geography with more than nine years of experience at Krishnagar Government College and Acharya B. N. Seal Government College. Presently he is a Professor of Geography in Sidho-Kanho-Birsha University, Purulia, West Bengal, India. He achieved National Scholarship and stood first position (First Class) at undergraduate level from Calcutta University. His areas of specialization are urban geography, environmental hazards and disaster analysis, GIS and remote sensing. He completed his Ph.D. on Urban Geography from the Department of Geography, University of Calcutta, West Bengal, India entitled "Emerging Urban Centres–Status and Possibilities: A Case Study of the District of North 24 Parganas, West Bengal." He has presented several research papers in international-, national-, and state-level seminars and conferences. He has published a number of articles in reputed peer-reviewed international and national journals and in edited books.

List of Figures

List of Tables

Chapter 1
Climatic Vulnerability and Migration: Evidence from Coastal Sundarbans

Abstract Coastal regions are particularly susceptible to climate-induced hazards, such as rising sea levels, storm surges, and coastal erosion, which threaten both human settlements and economic stability The Sundarbans, home to the world's largest contiguous mangrove forest along the Bay of Bengal, is among the most climate-vulnerable coastal regions. The increasing frequency of tropical cyclones in the Bay of Bengal is evident from the 121 cyclones recorded between 1980 and 2024 and the intensity or number or frequency of cyclone is highest near the Indian Sundarbans. Cyclones Aila (2009), Amphan (2020), and Yaas (2021) have caused mass displacement, with Cyclone Amphan alone impacting 60 million people and resulting in 80 fatalities in West Bengal. Agriculture and fisheries, the two principal livelihoods of the Sundarbans' coastal dwellers, are now under severe threat from saltwater intrusion, soil degradation, and storm damage. Consequently, many residents are compelled to migrate to other states in search of alternative employment opportunities. A nuanced understanding of these migration patterns can inform the formulation of adaptive policy strategies that enhance community resilience while mitigating displacement risks. Addressing climate-induced migration from coastal Sundarbans (India) requires a multi-pronged approach that integrates sustainable livelihood support, improved disaster preparedness, and inclusive policy frameworks to ensure the well-being and stability of affected populations. So, the particular nexus between climatic vulnerability and migration was tried to understand by a case study through an empirical database. The primary data was collected from 750 households which are categorized as migrant and non-migrant households.

Keywords Sundarbans · Cyclone · Migration · Displacement · Hazard · Pathar Pratima · Risk · Vulnerable

P. Mandal and S. Paul, *Climatic Vulnerability and Migration: Evidence from Indian Sundarbans*, Springer Geography, https://doi.org/10.1007/978-3-032-05858-4_1

1.1 Introduction

Climate change has emerged as one of the most pressing global challenges of the twenty-first century, exerting profound impacts on ecosystems, economies, and human populations. Rising global temperatures, shifts in precipitation patterns, increasing sea levels, and the growing frequency of extreme weather events collectively contribute to environmental degradation and economic instability (Adom et al. 2023; Han et al. 2024). According to the Intergovernmental Panel on Climate Change (IPCC), global warming disproportionately affects vulnerable regions by intensifying climatic hazards such as droughts, heat waves, floods, and tropical cyclones (IPCC 2014). These climatic disruptions have far-reaching consequences for livelihoods, food security, and human well-being. Smallholder farmers, particularly in climate-sensitive regions, face heightened uncertainties due to extreme weather conditions, land degradation, and technological deficits, which undermine agricultural productivity (Ahmad et al. 2022a, b). Furthermore, projections suggest that by 2050, an additional 78 million people will face hunger if climate change continues unabated (Maharjan et al. 2021). Given their limited access to resources and adaptive capacity, women, children, and marginalized communities remain among the most vulnerable groups confronting disproportionate climate risks (Yadav and Lal 2018).

Coastal regions are particularly susceptible to climate-induced hazards such as rising sea levels, storm surges, and coastal erosion, which threaten both human settlements and economic stability (Rabbani et al. 2022). In response to these challenges, migration has emerged as both an adaptation strategy and a coping mechanism for communities residing in climate hazard-prone areas (Mbiyozo 2020). Climate-induced migration refers to the movement of people driven by environmental changes, with sudden-onset events such as cyclones and floods, as well as slow-onset processes like sea-level rise and desertification, being key drivers (Zickgraf 2021). The International Organization for Migration (IOM) defines environmental migration as "the movement of people due to sudden or progressive changes in the environment that adversely affect their lives or living conditions" (IOM 2011). Climate-induced displacement is an urgent issue, as approximately two billion people globally have been affected by natural disasters over the past two decades, with many forced to migrate (Adom et al. 2023). Between 2008 and 2016, abrupt weather-related disasters displaced an average of 21.5 million people annually, while slow-onset climate stressors led to the migration of thousands more (UNHCR 2018).

Communities displaced due to climate change frequently encounter economic instability, challenges in social integration, and the erosion of cultural identity, thereby making climate-induced migration not only a humanitarian crisis but also a significant socio-economic challenge (Ahsan et al. 2022). Despite the increasing awareness of climate-induced displacement, policy responses remain inadequate and fragmented (Mittal et al. 2025). The UNFCCC Cancun Adaptation Framework (2010) acknowledges the necessity of implementing climate-resilient migration policies at national levels, yet these policies remain weak and insufficient (McLeman et al. 2021). There is a critical need to address existing policy gaps by examining the intricate linkages between climate vulnerability and migration at global, regional, and local scales, with particular emphasis on the Sundarbans region.

Rising climatic hazards are reshaping migration patterns globally, compelling people and communities to relocate from high-risk zones to safer areas (McLeman 2019). Extreme weather events such as sea-level rise, floods, droughts, and storms have driven substantial population movements (Rabbani et al. 2022). Environmental and climate-related disasters are projected to force more than 1.2 billion people to migrate globally by 2050 (Han et al. 2024). Regions with low adaptive capacity, economic vulnerabilities, and high geographic exposure to climate risks, such as Latin America, Africa, and South Asia, are expected to bear the highest burden (Birkmann et al. 2022). In coastal cities, rising sea levels and land subsidence have already displaced thousands, as evidenced in Lagos (Nigeria) and Jakarta (Indonesia), raising concerns about permanent population shifts (Osei-Amponsah and Abdulai 2025). The growing magnitude of climate-induced migration necessitates immediate global cooperation to develop sustainable adaptation and resettlement strategies, as it is poised to become a major geopolitical and policy concern in the near future (Liman et al. 2024).

South Asia is one of the most climate-vulnerable regions in the world, experiencing severe monsoons, cyclones, rising sea levels, and extreme temperatures (Srivastava 2020). Climate change exacerbates food and water insecurity and economic volatility, thereby triggering migration in countries such as India, Bangladesh, Pakistan, and Nepal (Kay et al. 2015). In India, agricultural productivity has declined due to extreme heat waves, snowstorms, and erratic rainfall patterns, resulting in large-scale displacement from rural areas (McMahon and Gray 2021; DasGupta et al. 2019). Similarly, Pakistan experienced massive climate-induced displacement in 2022 when catastrophic floods affected 33 million people and displaced 2.1 million individuals (Yousuf et al. 2023). In Bangladesh, flash floods in the Haor region impacted 4.2 million people (Bell et al. 2021; Rahman 2024). South Asia's economic growth is also threatened, as climate change is projected to reduce its GDP by 1.8–8.8% by 2050 (Ahmed 2014). The Internal Displacement Monitoring Centre (IDMC 2024) reports that floods and storms account for 52.76% and 45.98% of displacement cases in India, respectively. Despite these alarming trends, climate policies in South Asia remain inadequate in addressing migration as an adaptation strategy, leaving millions vulnerable to forced displacement (Islam and Khan 2020).

The Sundarbans, home to the world's largest contiguous mangrove forest along the Bay of Bengal, is among the most climate-vulnerable coastal regions. Climate-induced hazards such as rising sea levels, storm surges, agricultural land salinization, and frequent cyclones have severely affected local communities, compelling them to migrate (Ghosh and Mistri 2023). The increasing frequency of tropical cyclones in the Bay of Bengal is evident from the 121 cyclones recorded between 1971 and 2020 (Mondal et al. 2022). Cyclones Aila (2009), Amphan (2020), and Yaas (2021) have caused mass displacement, with Cyclone Amphan alone impacting 60 million people and resulting in 80 fatalities in West Bengal (OCHA 2020). Agriculture and fisheries, the two principal livelihoods of the Sundarbans' coastal dwellers, are now under severe threat from saltwater intrusion, soil degradation, and storm damage.

Consequently, many residents are compelled to migrate to other states in search of alternative employment opportunities (Brown and Nicholls 2015; Mandal 2018).

Given the region's heightened vulnerability, it is imperative to assess the complex dynamics of climate-induced migration in the Sundarbans. A nuanced understanding of these migration patterns can inform the formulation of adaptive policy strategies that enhance community resilience while mitigating displacement risks. Addressing climate-induced migration requires a multi-pronged approach that integrates sustainable livelihood support, improved disaster preparedness, and inclusive policy frameworks to ensure the well-being and stability of affected populations.

1.2 Evaluation of Research on Climate-Induced Migration

The research on climate-induced migration has expanded significantly in recent decades, reflecting the growing recognition of how environmental stressors influence human mobility. Scholars such as Sweijs et al. (2022) and Cabieses and Huerta-Montañez and Philipsborn (2024) have spearheaded interdisciplinary investigations, integrating socio-economic indicators with environmental data. This approach facilitates the development of simulation models and empirical studies that project future displacement trends and analyze the intricate relationship between climate variability and migration dynamics.

The World Bank Group (2018) established a robust framework for predicting migration flows under various climate scenarios, incorporating quantitative data from climate models, economic analyses, and demographic statistics. Similarly, Stave (2003) utilized a dynamic simulation model that incorporated climate variables, population health, and economic factors to generate valuable projections for policymakers. This methodology has been further extended through localized studies, such as those in Latin America, which demonstrate how environmental events like droughts, floods, and rising sea levels interact with socio-economic vulnerabilities to shape migration decisions (Almulhim et al. 2024).

Historical research has also contributed significantly to this field by contextualizing current migration trends. Case studies have illustrated how past civilizations adapted to environmental shifts, reinforcing the argument that climate has long served as a driver of human migration (Huntington et al. 2006). Understanding historical responses to climate variability allows for the refinement of contemporary models, improving their predictive accuracy in the context of ongoing global warming (McLeman 2019).

Despite these advancements, several critical gaps remain in the research. A major challenge lies in disentangling climate factors from other migration drivers (Borderon et al. 2019). Many studies acknowledge the nonlinear interactions among economic, political, and social variables, which are often oversimplified. For instance, models frequently assume a direct correlation between specific climatic events, such as droughts and migration. However, such models fail to capture the cumulative impact of multiple and overlapping stressors (Schewel et al. 2024).

Another significant limitation is the quality and availability of data on climate-induced migration. Global datasets provided by institutions such as the World Bank Group and the UN Environment Programme offer broad overviews but lack granularity in understanding localized migration patterns (Clemens and Kremer 2016). In highly vulnerable regions such as the Indian Sundarbans, data on climate-induced migration are either unavailable or inconsistently collected, thereby restricting the ability to monitor migration trends over time (Bandyopadhyay et al. 2023). Furthermore, discrepancies in data collection methodologies create additional barriers to accurately assessing migration flows (Cabieses and Huerta 2024).

A crucial area for future exploration is the role of political and institutional factors in shaping migration responses to climate change. Political conditions can either exacerbate or mitigate the impact of environmental stressors, necessitating further research in this domain to inform effective and comprehensive policy interventions (Islam et al. 2024). Addressing these research gaps will enhance the understanding of climate-induced migration, enabling policymakers to devise strategies that ensure both human security and environmental sustainability in vulnerable regions like the coastal Sundarbans.

1.3 Theoretical Background

Climatic vulnerability and migration have been intrinsically linked, particularly in environmentally sensitive regions such as the coastal Sundarbans. Migration, as conceptualized by Lee (1966) through the "push–pull" framework, is driven by various socio-economic and environmental factors that evolve spatially and temporally. Black et al. (2011) expanded upon this concept by categorizing migration drivers into five primary domains: economic, political, demographic, social, and environmental. These drivers seldom act in isolation; rather, their interplay dictates the magnitude and nature of migration, ranging from internal to international movements (Fig. 1.1).

Economic drivers are crucial in shaping migration patterns, as they encompass disparities in income, employment opportunities, and overall economic stability. In many developing nations, including India, economic growth disparities between rural and urban regions fuel internal migration, with individuals relocating in pursuit of better livelihoods. Conversely, economic downturns can also trigger short-term displacement, exacerbating socio-economic vulnerabilities (Black et al. 2011). Within the Sundarbans, economic instability—marked by fluctuating agricultural productivity and limited industrial growth—contributes to sustained out-migration, particularly among younger demographics seeking employment in urban centers.

Political drivers, including governance failures, policy frameworks, and security concerns, also significantly influence migration patterns. Instances of forced relocation due to infrastructure projects, land tenure insecurity, and political persecution can displace entire communities. Governance challenges in the Sundarbans, such as inadequate disaster response mechanisms and contested land rights, often compel

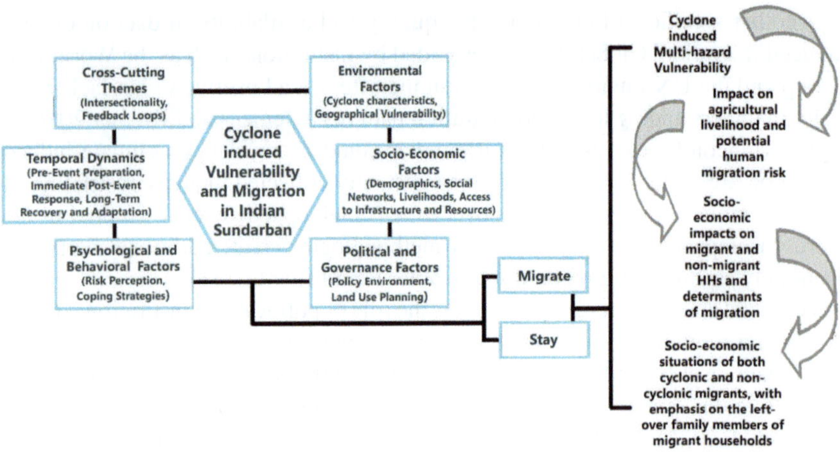

Fig. 1.1 Navigating migration complexities: the drivers conceptual model (adopted from Black et al. 2011)

populations to migrate as an adaptive response to socio-political instability (Black et al. 2011). Additionally, demographic drivers, such as population pressure and health risks, further shape migration flows. High population densities in the Sundarbans, coupled with inadequate healthcare facilities and recurrent disease outbreaks, exacerbate vulnerability, prompting migration to areas with better healthcare access and employment prospects.

Social factors, including familial obligations, cultural norms, and educational aspirations, add another layer of complexity to migration decisions. Migration for educational purposes has seen a global rise, with an estimated 2.8 million international student migrants in 2007, growing at an annual rate of 5.5% since 1999 (Dustmann and Glitz 2011). In the Sundarbans, limited access to quality education and skill development programs drives youth migration to urban centers, where opportunities for upward mobility are greater.

Environmental drivers, particularly climate-induced hazards, are among the most immediate triggers of displacement. The Sundarbans, being highly vulnerable to cyclones, floods, and coastal erosion, experience frequent environmental shocks that undermine traditional livelihoods, particularly agriculture and fisheries. Rapid-onset events such as Cyclone Amphan (2020) led to large-scale displacement, with many affected households unable to return due to habitat destruction and salinization of agricultural lands. Such environmental stressors exacerbate the pre-existing socio-economic vulnerabilities, compelling a continuous cycle of distress migration.

Given the multifaceted nature of migration drivers, it is imperative to adopt a holistic approach to climate adaptation and policy intervention. Understanding the interactions between economic, political, demographic, social, and environmental factors can aid in formulating resilient migration strategies. This study seeks to provide empirical insights into climate-induced migration patterns in the Sundarbans, thereby informing sustainable development and policy planning in the region.

1.4 Study Rationale and Significance

Tropical cyclones are among the most devastating meteorological phenomena globally, contributing to approximately 64% of all-natural disaster-related fatalities (Shen and Hwang 2019). The Bengal Delta, particularly the North Indian Ocean (NIO) Basin, is highly susceptible to cyclonic disturbances, accounting for nearly 7% of the world's cyclonic activity (Mondal et al. 2022). As part of this region, the Indian Sundarbans stands as one of the most ecologically fragile and socio-economically vulnerable deltaic systems, protected by an intricate network of mangroves and a complex drainage system (Ghosh and Mistri 2023). The Bay of Bengal's rising sea surface temperatures, frequently reaching 30–32 °C in the pre-monsoon period, contribute to frequent and intense tropical cyclones (Kumar et al. 2024; Ghosh and Mistri 2023).

Over the past five decades, the Bay of Bengal has experienced 121 cyclonic events, significantly disrupting coastal agriculture due to saline water intrusion and soil salinization (Ghosh and Mistri 2023). This dynamic interplay between environmental hazards and human vulnerability necessitates an in-depth examination of the link between climatic vulnerability and migration, particularly in regions like the Sundarbans where livelihoods are predominantly dependent on climate-sensitive sectors such as agriculture and fisheries (Pourghasemi et al. 2020a, b). According to the Internal Displacement Monitoring Centre dataset, there is an average displacement of around 170,000 individuals per event, with a median displacement of approximately 1,500 people. The most extreme recorded displacement event affected up to 6.9 million individuals. Among the various types of natural hazards, floods emerge as the most frequent and consequential, often leading to displacements of over 2 million people in some instances. The severity of storms, particularly cyclones, is evident as they precipitate significant population movements. These events underscore such natural phenomena of extreme intensity and destructive potential (Fig. 1.2).

Given that climate-induced displacement is a growing concern, the research aims to explore the complexities of migration patterns triggered by cyclonic hazards. Migration, in this context, is not merely a reactive measure to immediate displacement but also a long-term adaptive strategy to worsening environmental conditions (Crnčević and Orlović 2018). Understanding these processes can inform targeted policies that address both immediate disaster relief and long-term adaptation strategies.

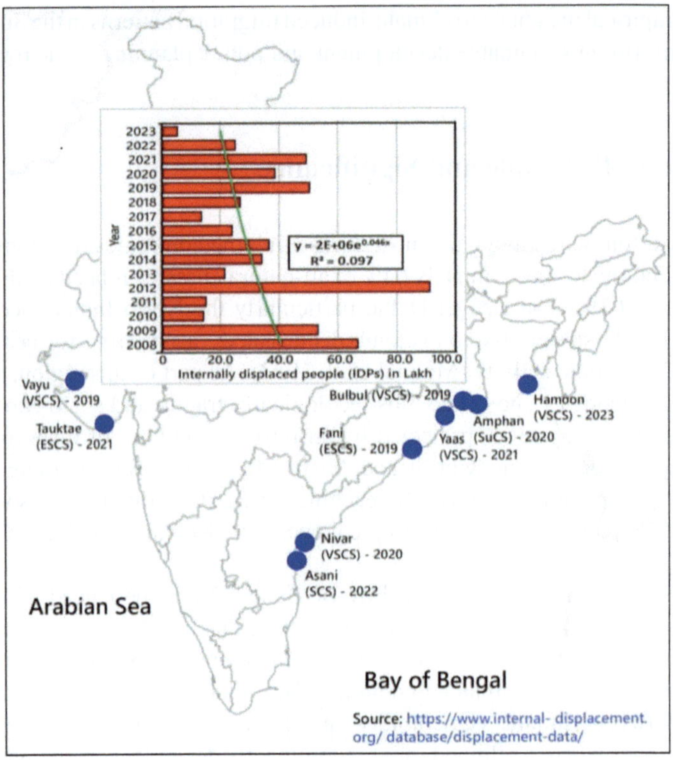

Fig. 1.2 Internally displaced population (IDPs) in India from 2008 to 2023 and landfalls of intense cyclones in the last five years

1.5 Significance of the Study

1.5.1 Understanding the Nexus Between Climate Vulnerability and Migration

Cyclones have profound socio-economic and demographic implications. They lead to loss of housing, destruction of livelihoods, and extensive damage to infrastructure, forcing affected populations to migrate either temporarily or permanently (Palinkas 2020). Inhabitants of the Sundarbans frequently relocate due to a combination of economic, environmental, and social stressors exacerbated by repeated cyclone-induced devastation (Pogue 2021). This study aims to analyze the multi-dimensional vulnerability induced by climate change and how it manifests in migration decisions, thereby providing an empirical basis for future mitigation and adaptation efforts.

1.6 Regional Relevance: The Case of the Indian Sundarbans

The Indian Sundarbans is a high-risk zone, experiencing roughly nine cyclonic events per decade, a third of which are classified as severe (Ghosh and Roy 2022). Historical records indicate that cyclones such as Sidr (2007), Aila (2009), Phailin (2013), Hudhud (2014), Fani (2019), Bulbul (2019), Amphan (2020), and Yaas (2021) (Fig. 1.3) have significantly disrupted local livelihoods. More than 60% of the population in the Sundarbans relies on agriculture, a sector highly susceptible to climate-induced shocks such as saline water intrusion and land degradation (Sen and Ghorai 2019). Given that storm surges during cyclonic events often exceed embankment heights, large tracts of agricultural land become unsuitable for cultivation, thereby accelerating rural distress and out-migration (Hossain 2015; Islam et al. 2024). The region's cropping intensity remains considerably lower than the state average, further exacerbating food insecurity and economic instability (Mandal 2018). This study will investigate these climatic stressors in detail to understand their cascading effects on migration patterns. By analyzing how climate stress influences mobility, policymakers can better tailor intervention strategies to mitigate displacement risks.

The Indian Sundarbans, located in the deltaic region of West Bengal, is a globally recognized biodiversity hotspot and UNESCO World Heritage Site, yet it faces severe climatic vulnerabilities that profoundly impact its socio-economic fabric. Pathar Pratima, a community development block in South 24 Parganas, West Bengal,

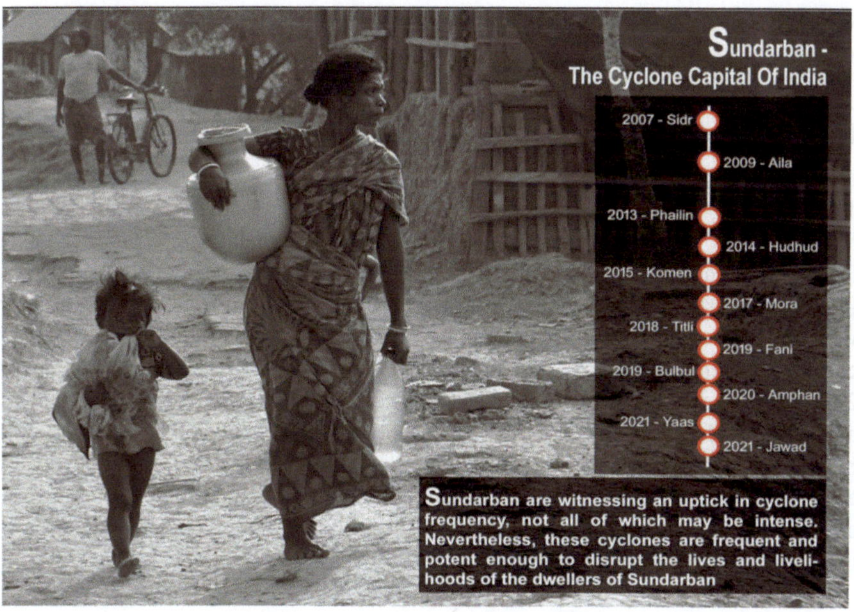

Fig. 1.3 Series of cyclonic storm hit Sundarbans in a regular interval

was selected as a case study for its representativeness of these challenges, making it highly relevant for research on climatic vulnerability and human mobility. This block is characterized by its low-lying coastal geography, with an average elevation of 2.8–3.5 m above mean sea level, rendering it acutely susceptible to frequent cyclonic storms, storm surges, and sea-level rise, which have intensified due to climate change. Notably, cyclones like Aila (2009), Bulbul (2019), and Amphan (2020) have caused significant embankment breaching and saline water intrusion, severely disrupting agriculture, the primary livelihood for 77.55% of the local economy. The socio-economic backwardness of Pathar Pratima, marked by high poverty levels (with 47.5% of households landless and 41% owning less than one acre) and limited access to modern energy and irrigation, exacerbates its vulnerability. These conditions have driven significant out-migration, as residents seek alternative livelihoods due to dwindling agricultural productivity and resource depletion. Furthermore, research highlights Pathar Pratima's high socio-economic vulnerability index (SeVI), attributed to constant exposure to cyclones, fragile infrastructure, and impoverished social structures, making it a critical site for studying adaptation and migration dynamics. The block's proximity to the Bay of Bengal amplifies its exposure to environmental hazards, with studies noting a sea-level rise rate of 3.14 mm per year, higher than the global average, contributing to land erosion and salinization. This environmental stress, coupled with limited adaptive capacity due to poor governance and inadequate disaster management, underscores Pathar Pratima's relevance as a case study. By focusing on this block, the research can elucidate the interplay between climatic stressors and human mobility, offering insights into sustainable adaptation strategies for deltaic communities globally, while addressing the urgent need to integrate local knowledge and resilience-building measures into policy frameworks.

1.7 Contributions to Policy and Sustainable Development

The study aligns with international climate adaptation and disaster risk reduction strategies, particularly those outlined in the Sustainable Development Goals (SDGs). Climate-induced displacement has been recognized as a critical challenge in global governance, yet regional-specific studies remain insufficient. This research will provide localized empirical insights that contribute to global discourses on climate migration and sustainable development (Reuveny 2021; Cabieses and Huerta 2024). The findings of this study can serve as a knowledge base for policy planners, aiding in the development of comprehensive migration policies that integrate both humanitarian and long-term resilience-building approaches. Policies focusing on adaptive strategies, such as improved embankment systems, climate-resilient agriculture, and alternative livelihoods, can be informed by a granular understanding of the region-specific climate migration nexus.

1.8 Bridging Research Gaps in Climate-Induced Migration

Despite extensive discussions on climate change and migration, research often isolates environmental drivers from socio-economic and demographic variables (Huntington et al. 2006). Existing studies largely focus on singular climatic factors such as flooding or droughts while neglecting the compounded effects of multiple hazards on vulnerable communities. This study seeks to fill that gap by providing a comprehensive analysis that integrates climate change, socio-economic vulnerabilities, and migration trends in the Sundarbans.

Localized studies remain crucial to developing holistic adaptation and mitigation frameworks. According to the World Bank Group (2018) and the UN Environment Programme (2024), millions are at risk of displacement due to climate-induced hazards, yet policy responses often lack the granularity required for effective intervention. This research will bridge that gap by offering a region-specific understanding of climate-induced migration, providing a roadmap for integrating disaster preparedness with sustainable resettlement policies.

This study will contribute significantly to the ongoing discourse on climate vulnerability and migration by focusing on the Indian Sundarbans, a region highly susceptible to cyclonic hazards. By examining how climate-induced vulnerabilities drive migration patterns, this research aims to provide empirical evidence to support policy formulation and disaster risk management strategies. Understanding migration as both an adaptive and distress-driven response to environmental changes will enable stakeholders to devise sustainable interventions that enhance community resilience. In the broader context, this study will not only enrich scholarly literature but also provide actionable insights to policymakers working to mitigate climate-induced displacement in coastal regions worldwide.

1.9 Statement of the Research Problem

Climate-induced migration has emerged as a critical issue, particularly in the context of coastal regions that are highly vulnerable to extreme weather events such as cyclones. The Sundarbans, a low-lying coastal region, faces recurrent cyclonic hazards that result in widespread destruction, displacement, and socio-economic instability. This research investigates the intricate relationship between climatic vulnerability and migration, focusing on cyclone-induced displacement in the Sundarbans.

Cyclones exacerbate existing vulnerabilities by causing extensive damage to infrastructure, loss of livelihoods, and disruption of essential services such as food, water, and healthcare. Vulnerability in this context is multifaceted, encompassing socio-economic, environmental, and demographic dimensions. Socio-economic vulnerability stems from poverty, inadequate resources, and limited access to adaptive mechanisms (Islam 2021). Environmentally, the proximity of coastal communities to high-risk zones, coupled with ecosystem degradation, heightens susceptibility to cyclonic events (Mishra et al. 2024). Demographic factors, including age, gender, and household composition, further influence the capacity of individuals and communities to withstand and recover from such disasters (Adger et al. 2019).

Migration serves as both an adaptive strategy and a forced response to cyclonic hazards. Immediate evacuations occur to escape direct threats posed by cyclones, while longer term migration results from the destruction of homes, farmlands, and fishing resources, compelling affected populations to seek alternative livelihoods elsewhere. Understanding the drivers, patterns, and decision-making processes of cyclone-induced migration is crucial for designing effective response strategies. The dynamics of migration are shaped by push factors such as economic instability, land degradation, and recurring disasters, as well as pull factors like employment opportunities and social networks in destination areas (McLeman et al. 2021).

Moreover, migration has profound and enduring implications for both the areas of origin and destination. In origin areas, out-migration can lead to demographic shifts, labor shortages, and changes in social structures, potentially exacerbating economic decline and environmental degradation (Roy et al. 2022). Conversely, in destination areas, the influx of migrants may strain existing resources, infrastructure, and services, leading to socio-economic tensions and challenges in integration (Kartiki 2011). Therefore, long-term studies are essential to assess how cyclone-induced migration reshapes the socio-economic and environmental landscape over time, as well as to identify challenges and opportunities for adaptation and development (Fig. 1.4).

A key aspect of this research involves refining methodologies to comprehensively assess cyclone-induced vulnerability and migration. There is a pressing need for improved analytical frameworks, data collection techniques, and modeling approaches that can effectively capture the complex interactions between cyclonic hazards, vulnerability factors, and migration dynamics (Ghosh and Mistri 2023). Spatially explicit assessments, using geospatial tools and empirical data, are critical for identifying high-risk populations, assessing migration trends, and predicting future displacement scenarios in cyclone-prone regions (DasGupta et al. 2019).

Overall, this study underscores the multi-dimensional nature of cyclone-induced vulnerability and migration in the Sundarbans. By integrating socio-economic, environmental, and spatial analyses, the research aims to enhance understanding of the mechanisms driving climate-induced displacement and inform policy interventions that can strengthen resilience and adaptive capacities in cyclone-affected communities. Addressing this issue is imperative for developing sustainable and inclusive strategies that mitigate risks, reduce vulnerability, and promote climate-resilient development pathways in coastal regions.

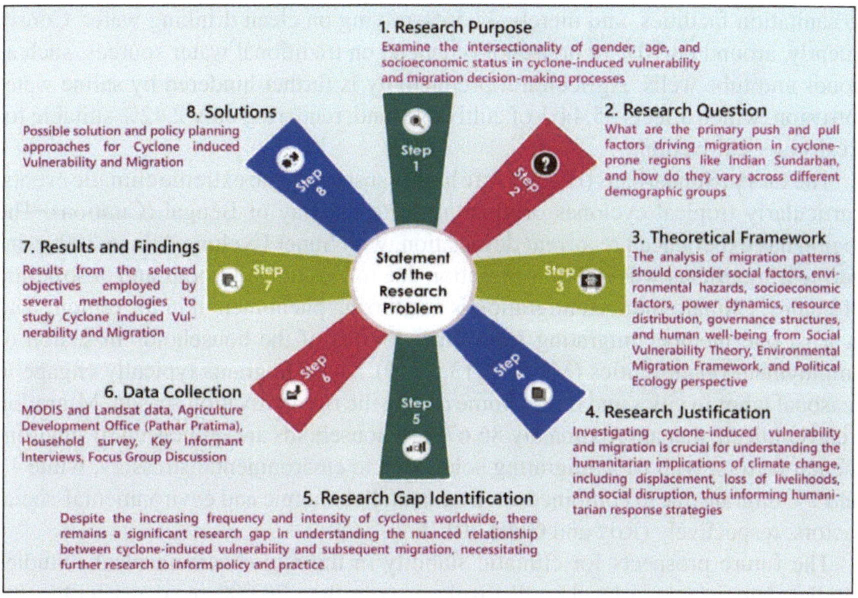

Fig. 1.4 Statement of research problem in a schematic overview

1.10 Materials and Methods

1.10.1 Study Area

The Indian Sundarbans, a critical ecological region situated along the western periphery of the Ganges–Brahmaputra–Meghna (GBM) Delta, is particularly vulnerable to climate-induced displacement. This region comprising with 103 low-lying deltaic islands accounts for 19% of the total Sundarbans area, while the remaining 81% falls within Bangladesh. Administratively, the Indian Sundarbans span across North and South 24 Parganas districts in the state of West Bengal, encompassing 19 blocks and 4 subdivisions. The region is home to a population of 4,426,259 individuals residing in 955,040 households, with a sex ratio of 956 females per 1,000 males. The literacy rate stands at 64.30%, though disparities in educational access persist due to geographical isolation and economic hardships (Census of India 2011).

The socio-economic framework of the Indian Sundarbans is predominantly agrarian. Approximately 35.56% of the population is engaged in agriculture, primarily paddy cultivation and betel leaf farming, while fishing and aquaculture sustain 4.79% of the population. Other livelihood activities include honey collection, handicrafts, and daily wage labor. However, a significant portion of the population, approximately 70.32%, falls under the Below Poverty Line (BPL) category, highlighting the socio-economic vulnerability of the region (Census of India 2011). Infrastructure remains inadequate, with only 58% of the population having access to sanitation facilities, and merely 37.55% relying on clean drinking water. Consequently, around 68.70% of households depend on traditional water sources, such as ponds and tube wells. Agricultural productivity is further hindered by saline water intrusion, which affects 35.44% of cultivable land, rendering only 2.42% suitable for freshwater cultivation.

The Indian Sundarbans (Fig. 1.5) are highly susceptible to extreme climatic events, particularly tropical cyclones originating from the Bay of Bengal (Citation). The region has experienced recurrent devastation, with Super Cyclone Aila in 2009 triggering a significant wave of out-migration due to loss of livelihoods and destruction of homes. Climate-induced migration is a recurring phenomenon in the region, with at least one member migrating from three-fourths of the households in search of employment opportunities (Mistri 2013, 2019). These migrants typically engage in seasonal labor in cities and return home during the rice cultivation season. Migration trends indicate that approximately 86.67% of households are influenced by environmental factors, with 61% migrating solely due to environmental stressors, while 17 and 2% migrate due to combined environmental-economic and environmental-social factors, respectively (Roy and Guha 2017).

The future prospects for climatic stability in the region remain bleak. Studies predict that rising sea levels will displace more than 70,000 residents within the next decade, with Sagar Island alone projected to lose 30,000 inhabitants (Hazra et al. 2002; Roy and Guha 2017). Limited local infrastructure, inadequate government support, and constrained asset holdings further exacerbate out-migration trends,

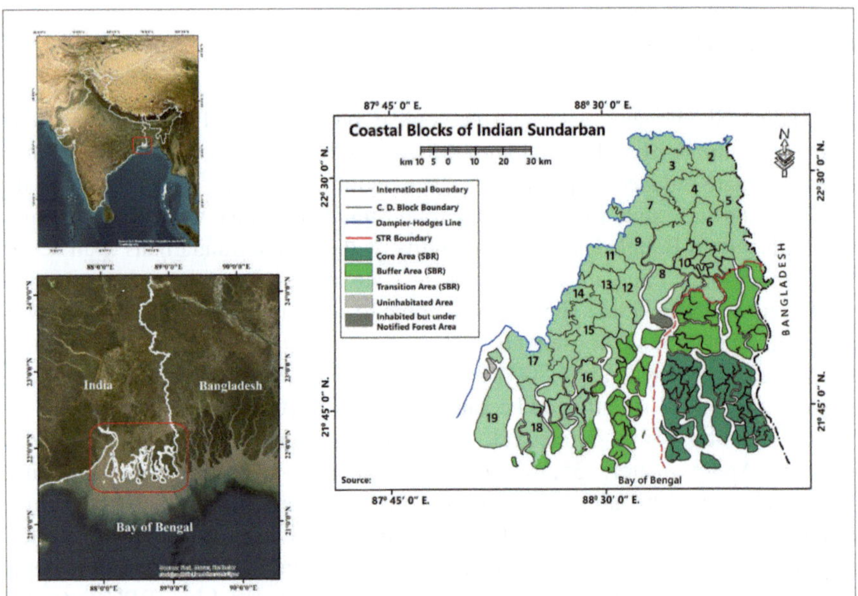

Fig. 1.5 Geographical setting of the study area

Fig. 1.6 Livelihood on the edge: a couple navigates the waters of the Sundarbans, a climate hotspot where fishing is both a way of life and a survival strategy amid increasing environmental uncertainty. *Source* By authors on 19-04-2023

forcing residents, particularly male members, to migrate to Kerala, Mumbai, Gujarat, and Odisha in search of better livelihoods (Mistri 2013). Migration has emerged as an inevitable survival strategy, particularly for landless and smallholder households, who face repeated crop failures due to erratic climatic patterns. Mistri and Das (2019) assert that male out-migration will continue to shape the economic structure of the region, with remittances playing a crucial role in household sustenance. Skilled and experienced migrants often secure better wages, resulting in wealth generation and economic resilience in their home communities.

Given the region's acute climatic vulnerability (Fig. 1.6) and socio-economic fragility, the Indian Sundarbans remain a critical area for studying the nexus between climate change and migration. The persistent threats of coastal inundation, erosion, and extreme weather events underscore the urgent need for strategic adaptation measures and policy interventions to mitigate displacement risks and enhance livelihood security for the region's inhabitants.

1.11 Conclusion

So, the introduction chapter has successfully established the background of the research. The whole research was justified in world, regional (South Asia), and local perspective (Indian Sundarbans). Here it also established that why the issue is more important and challenges for the research scholar in world as well as local perspective like as environmental hazards hotspot region as Sundarbans (India). Finally revealed the significance of the research in Sundarbans which is one of the most vulnerable areas due to cyclone intensity and also we analyzed the cyclone intensity from 1980 to 2024 and represented it by a map (Fig. 1.7).

The research findings will explain the migration pattern, such as the migrant's age, income, gender inequality, social background, and economic class. The study also revealed that the people from Sundarbans migrated temporarily. They did not migrate with family due to expenditure in the destination place. So, they send remittances on a monthly basis or one time in 3 months, and the decision to spend their money also depends on their family structure. The father or mother takes it for the joint family, but in the case of a nuclear family, it is taken by the wife or migrant himself. The cyclone-risk-prone region highly influences to decide to migrate compared to the low vulnerable region like near the coast where people have migrated more compared to people living in inland areas. From the findings, the study concluded that migration could be an adaptation strategy for coastal dwellers to cope with the adverse effects of cyclonic hazards as they increase their resilience and overall quality of life after migration. So, here, the government's role is to make plans for voluntary migrants, not only to prevent them but also to give them the security of their left-behind families at the origin place and in the destination location, particularly the migrant workers. The findings of this study can serve as a knowledge base for policy planners, aiding in the development of comprehensive migration policies that integrate humanitarian and long-term resilience-building approaches. Policies focusing on adaptive strategies, such as improved embankment systems, climate-resilient agriculture, and alternative livelihoods, can be informed by a granular understanding of the region-specific climate migration nexus.

Before going to formulate the specific objective of the whole research, we have tried to find out the research gap through bibliometric analysis in the next chapter.

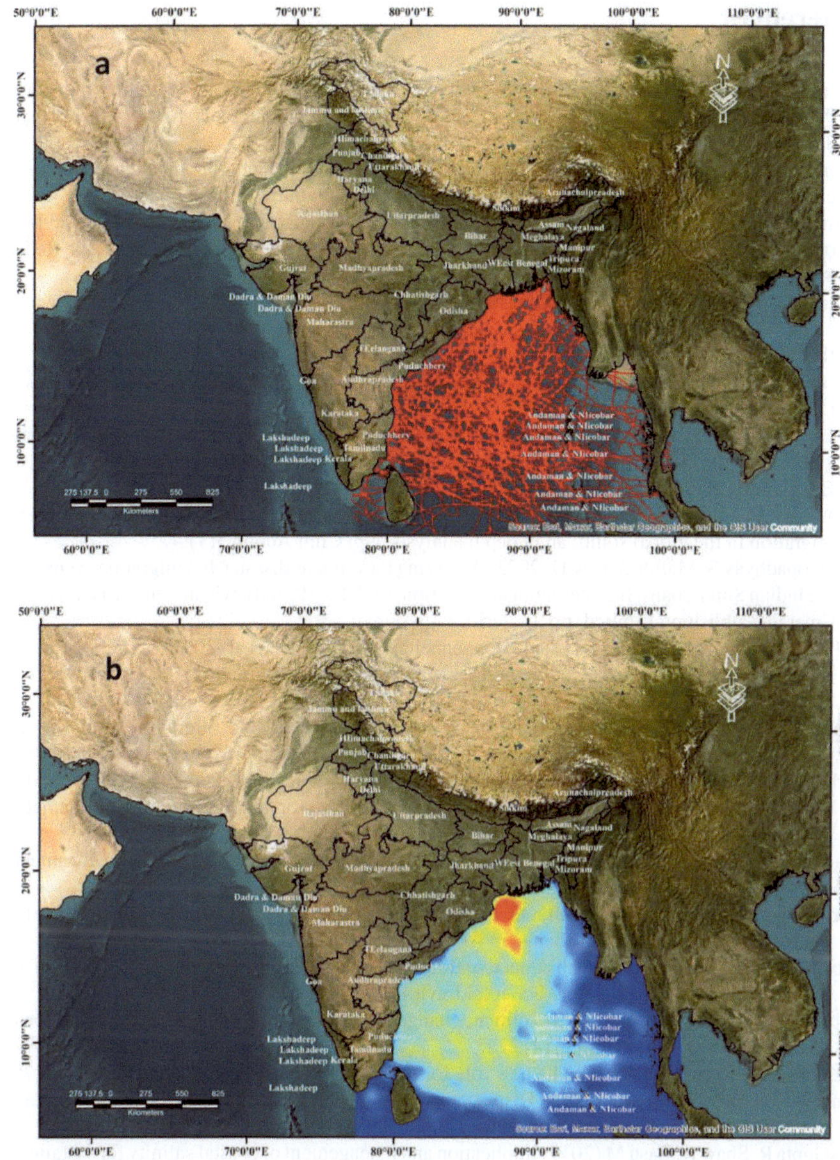

Fig. 1.7 Special view of **a** cyclonic track and **b** intensity of it through density analysis in Bay of Bengal from 1980 to 2024

References

Adger WN, Boyd E, Fábos A, Fransen S, Jolivet D, Neville G et al (2019) Migration transforms the conditions for the achievement of the sustainable development goals. Lancet Planet Health 3(11):e440–e442

Adom RK, Simatele MD, Das DK (2023) Unravelling the social and economic consequences of climate change on sustainable livelihoods on coastal communities in South Africa

Ahmad MM, Yaseen M, Saqib SE (2022a) Climate change impacts of drought on the livelihood of dryland smallholders: implications of adaptation challenges. Int J Disaster Risk Reduct 80:103210

Ahmad M, Ahmed Z, Yang X, Hussain N, Sinha A (2022b) Financial development and environmental degradation: do human capital and institutional quality make a difference? Gondwana Res 105:299–310

Ahmed M (2014) Assessing the costs of climate change and adaptation in South Asia. Asian Development Bank

Ahsan MN, Khatun F, Kumar P, Dasgupta R, Johnson BA, Shaw R (2022) Promise, premise, and reality: the case of voluntary environmental non-migration despite climate risks in coastal Bangladesh. Reg Environ Chang 22(1):1

Almulhim AI, Alverio GN, Sharifi A, Shaw R, Huq S, Mahmud MJ et al (2024) Climate-induced migration in the global south: an in depth analysis. Npj Clim Action 3(1):47

Bandyopadhyay S, Mallik C, Roy U (2023) Assessing the climate-disaster-led migration scenario in the Indian Sundarbans. In: International migration, COVID-19, and environmental sustainability. Emerald Publishing Limited, pp 97–115

Bell AR, Wrathall DJ, Mueller V, Chen J, Oppenheimer M, Hauer M et al (2021) Migration towards Bangladesh coastlines projected to increase with sea-level rise through 2100. Environ Res Lett 16(2):024045

Biella R, Hoffmann R, Upadhyay H (2022) Climate, agriculture, and migration: exploring the vulnerability and outmigration nexus in the Indian Himalayan region. Mt Res Dev 42(2):R9–R21

Birkmann J, Jamshed A, McMillan JM, Feldmeyer D, Totin E, Solecki W et al (2022) Understanding human vulnerability to climate change: a global perspective on index validation for adaptation planning. Sci Total Environ 803:150065

Black R, Adger WN, Arnell NW, Dercon S, Geddes A, Thomas D (2011) The effect of environmental change on human migration. Glob Environ Chang 21:S3–S11

Borderon M, Sakdapolrak P, Muttarak R, Kebede E, Pagogna R, Sporer E (2019) Migration influenced by environmental change in Africa. Demogr Res 41:491–544

Brown S, Nicholls RJ (2015) Subsidence and human influences in mega deltas: the case of the Ganges–Brahmaputra–Meghna. Sci Total Environ 527:362–374

Cabieses B, Huerta C (2024) Effects of climate change on migration in Latin America and Caribbean: a scoping review. Front Clim 6:1412285

Clemens MA, Kremer M (2016) The new role for the World Bank. J Econ Perspect 30(1):53–76

Crnčević T, Orlović Lovren V (2018) Displacement and climate change: improving planning policy and increasing community resilience. Int J Clim Chang Strat Manag 10(1):105–120

DasGupta R, Shaw R, Basu M (2019) Implication and management of coastal salinity for sustainable community livelihood: case study from the Indian Sundarban Delta. In: Coastal management. Academic Press, pp 251–269

Dustmann C, Glitz A (2011) Migration and education. In: Handbook of the economics of education, vol 4. Elsevier, pp 327–439

Ghosh S, Mistri B (2023) Cyclone-induced coastal vulnerability, livelihood challenges and mitigation measures of Matla-Bidya inter-estuarine area, Indian Sundarban. Nat Hazards 116(3):3857–3878

Ghosh S, Roy S (2022) Climate change, ecological stress and livelihood choices in Indian Sundarban. In: Climate change and community resilience, vol 399

Han Q, Kumar R, Kumar A (2024) Climate change and human migration: perspectives for environmentally sustainable societies. J Geochem Explor 256:107352

Hazra S, Ghosh T, DasGupta R, Sen G (2002) Sea level and associated changes in the Sundarbans. Sci Cult 68(9/12):309–321

https://www.internal-displacement.org/global-report/grid2024/

https://www.unep.org/resources/annual-report-2024

https://www.censusindia.gov.in/nada/index.php/catalog/1362

Hossain MN (2015) Analysis of human vulnerability to cyclones and storm surges based on influencing physical and socioeconomic factors: evidences from coastal Bangladesh. Int J Disaster Risk Reduct 13:66–75

Huerta-Montañez G, Philipsborn R (2024) The trauma of leaving all behind: climate and environmental migration and children. Acad Pediatr 24(5):50–52

Huntington KW, Blythe AE, Hodges KV (2006) Climate change and Late Pliocene acceleration of erosion in the Himalaya. Earth Planet Sci Lett 252(1-2):107–118

International Organization for Migration (2011) World migration report 2011: communicating effectively about migration. International Organization for Migration. https://www.publications.iom.int/books/world-migration-report-2011

IPCC (2014). Climate change 2014: synthesis report. In: Core Writing Team, Pachauri RK, Meyer LA (eds) Contribution of working groups I, II and III to the fifth assessment report of the intergovernmental panel on climate change. IPCC, Geneva, Switzerland, p 151

Islam MR, Khan NA (2020) Threats, vulnerability, resilience and displacement among the climate change and natural disaster-affected people in South-East Asia: an overview. In: Climate change mitigation and sustainable development, pp 111–138

Islam MS, Lee Z, Shaleh A, Soo HS (2024) The United Nations environment assembly resolution to end plastic pollution: challenges to effective policy interventions. Environ Dev Sustain 26(5):10927–10944

Islam SLU (2021) Assessment of the potential of bioengineering intervention for retarding cyclonic wind and storm surge in coastal Bangladesh

Kartiki K (2011) Climate change and migration: a case study from rural Bangladesh. Gend Dev 19(1):23–38

Kay S, Caesar J, Wolf J, Bricheno L, Nicholls RJ, Islam AS et al (2015) Modelling the increased frequency of extreme sea levels in the Ganges–Brahmaputra–Meghna delta due to sea level rise and other effects of climate change. Environ Sci Process Impacts 17(7):1311–1322

Kumar A, Chakraborty A, Sagar AK (2024) Analysis of the Bay of Bengal cyclone and its rejuvenation in the Arabian Sea after passing over peninsular India. Geomat Nat Haz Risk 15(1):2433110

Lee ES (1966) A theory of migration. Demography 3:47–57

Liman M, Nuhu Z, Halliru SL, Sodangi AB (2024) Migration as an adaptation strategy to climate change and conflict stresses: exemplifying north-western Nigeria. In: Climate change and regional socio-economic systems in the global south: resilience strategies for sustainable development. Springer Nature Singapore, Singapore, pp 233–248

Maharjan A, Tuladhar S, Hussain A, Mishra A, Bhadwal S, Ishaq S et al (2021) Can labour migration help households adapt to climate change? Evidence from four river basins in South Asia. Climate Dev 13(10):879–894

Mandal GC (2018) The role of agricultural diversification in rural development: a case study of mountain livelihood systems in the himalayan region of West Bengal. Doctoral dissertation, University of North Bengal

Mbiyozo AN (2020) Migration: a critical climate change resilience strategy. Institute for Security Studies

McLeman R (2019) International migration and climate adaptation in an era of hardening borders. Nat Clim Chang 9(12):911–918

McLeman R, Wrathall D, Gilmore E, Thornton P, Adams H, Gemenne F (2021) Conceptual framing to link climate risk assessments and climate-migration scholarship. Clim Chang 165:1–7

McMahon K, Gray C (2021) Climate change, social vulnerability and child nutrition in South Asia. Glob Environ Chang 71:102414
Mishra M, Guria R, Paul S, Baraj B, Santos CAG, Dos Santos CAC, da Silva RM (2024) Geo-ecological, shoreline dynamic, and flooding impacts of Cyclonic Storm Mocha: a geospatial analysis. Sci Total Environ 917:170230
Mistri A (2013) Migration and sustainable livelihoods: a study from Sundarban biosphere reserve. Asia Pac J Soc Sci 5(2):76–102
Mistri A (2019) Is the migration from Indian Sundarban an environmental migration? Investigating through sustainable livelihood approach (SLA). Asian Profile 47(3):195–219
Mistri A, Das B (2019) The Sundarban and its environment. In: Environmental Change, Livelihood Issues and Migration: Sundarban Biosphere Reserve, India. Singapore: Springer Singapore, pp 21–40
Mittal S, Chandel A, Le TM (2025) Social and economic impacts of climate-induced migration and displacement. In: Effects of climate change on social and economic factors. IGI Global, pp 49–82
Mondal M, Biswas A, Haldar S, Mandal S, Bhattacharya S, Paul S (2022) Spatio-temporal behaviours of tropical cyclones over the bay of Bengal Basin in last five decades. Trop Cyclone Res Rev 11(1):1–15
Osei-Amponsah C, Abdulai IA (2025) Co-creating climate future pathways for northwestern Ghana: the use of the three horizons framework. Futures 103567
Palinkas LA (2020) Global climate change, population displacement, and public health. Springer
Pogue D (2021) How to prepare for climate change: a practical guide to surviving the chaos. Simon and Schuster
Pourghasemi HR, Gayen A, Edalat M, Zarafshar M, Tiefenbacher JP (2020a) Is multi-hazard mapping effective in assessing natural hazards and integrated watershed management? Geosci Front 11(4):1203–1217
Pourghasemi HR, Kariminejad N, Amiri M, Edalat M, Zarafshar M, Blaschke T, Cerda A (2020b) Assessing and mapping multi-hazard risk susceptibility using a machine learning technique. Sci Rep 10(1):3203
Rabbani MG, Cotton M, Friend R (2022) Climate change and non-migration—exploring the role of place relations in rural and coastal Bangladesh. Popul Environ 44(1):99–122
Reuveny R (2021) Climate-related migration and population health: social science-oriented dynamic simulation model. BMC Public Health 21:1–22
Roy C, Guha I (2017) Economics of climate change in the Indian Sundarbans. Glob Bus Rev 18(2):493–508
Roy P, Pal SC, Chakrabortty R, Islam ARMT, Chowdhuri I, Saha A (2022) Impact of ineffective measures on the increasing land degradation in a monsoon-dominated region of India: issues and policy implications. Land Degrad Dev 33(16):3174–3185
Schewel K, Dickerson S, Madson B, Nagle Alverio G (2024) How well can we predict climate migration? A review of forecasting models. Front Clim 5:1189125
Sen HS, Ghorai D (2019) The Sundarbans: a flight into the wilderness. In: The Sundarbans: a disaster-prone eco-region: increasing livelihood security, pp 3–28
Shen G, Hwang SN (2019) Spatial–temporal snapshots of global natural disaster impacts revealed from EM-DAT for 1900–2015. Geomat, Nat Hazards Risk
Srivastava RK (2020) Manifestations of climate change in South Asia. In: Managing urbanization, climate change and disasters in South Asia. Singapore: Springer Singapore, pp 135–198
Stave KA (2003) A system dynamics model to facilitate public understanding of water management options in Las Vegas, Nevada. J Environ Manag 67(4): 303–313
Sweijs T, De Haan M, Van Manen H (2022) Unpacking the climate security nexus: seven pathologies linking climate change to violent conflict. Hague Centre for Strategic Studies
United Nations High Commissioner for Refugees (2018) UNHCR global report 2017. United Nations High Commissioner for Refugees. https://www.unhcr.org/media/unhcr-global-report-2017

United Nations Office for the Coordination of Humanitarian Affairs (2021) OCHA annual report 2020. United Nations Office for the Coordination of Humanitarian Affairs. https://www.unocha.org/publications/report/world/ocha-annual-report-2020
United Nations Office for the Coordination of Humanitarian Affairs (OCHA) (2020, May 26)
World Bank Group (2018) Women, business and the law 2018. World Bank Publications
Yadav SS, Lal R (2018) Vulnerability of women to climate change in arid and semi-arid regions: the case of India and South Asia. J Arid Environ 149:4–17
Yousuf J, Mehmood H, Aquil S, Rija A, Rahmat ZS, Malikzai A (2023) Effects of floods on the mental health of Pakistanis: a commentary. Ann Med Surg 85(5):2253–2255
Zickgraf C (2021) Climate change, slow onset events and human mobility: reviewing the evidence. Curr Opin Environ Sustain 50:21–30

Chapter 2
Understanding Climate-Induced Migration Through Bibliometric Analysis: Vulnerabilities, Patterns, and Regional Insights

Abstract Climate-induced migration has become a significant global issue, particularly concerning labor migration, poverty, and water-related hazards. This study analyzes the growing body of literature on climate-induced migration, focusing on rural–urban migration, agricultural practices, and migration as an adaptation strategy. The primary objective is to identify key themes, patterns, and influential works within the field through bibliometric analysis using Scopus and Web of Science databases. The analysis spans publications from 2007 to 2023, revealing a shift from theoretical discussions to empirical studies and quantitative methods. The United States, China, the UK, and France have been the leading contributors to this research. Core themes include climate change, migration, and adaptive management, particularly in relation to cyclone-induced migration. The study also highlights localized research in regions like Khulna and Sub-Saharan Africa, emphasizing the vulnerability and socio-economic impacts of climate change on migration patterns. Despite these insights, such localized findings have yet to be fully integrated into broader global discussions. The Sundarbans, a region facing shared cyclonic hazards across India and Bangladesh, serves as a focal point for the study, underscoring the need for collaborative solutions to address the climate-induced migration crisis. The findings suggest that while climate-induced migration has been studied across various contexts, there is still a critical need for localized insights to be incorporated into the larger global migration discourse. Further research is recommended to explore the mechanisms of adaptive management and to integrate regional studies into global migration frameworks for more effective solutions.

Keywords Climate-induced migrations · Hazards · Vulnerability · Bibliometric · Synthesize

P. Mandal and S. Paul, *Climatic Vulnerability and Migration: Evidence from Indian Sundarbans*, Springer Geography, https://doi.org/10.1007/978-3-032-05858-4_2

2.1 Introduction

The issue of climate-induced migration has garnered significant attention from scholars, researchers, and politicians worldwide in the past few decades (Milan-García et al. 2021; Ghosh 2021). Climate migration research began with few publications, focusing on labor migration, poverty, and water-related hazards (Bamel et al. 2020). The research methods were largely qualitative, using case studies and interviews to capture the complex experiences of individuals and communities facing environmental stressors (Adams 2016; Khalil et al. 2021; Mallick et al. 2021). This period saw a significant expansion in the volume and scope of climate migration research, driven by increased global awareness of climate change and its potential impacts (Mandal et al. 2022). The Intergovernmental Panel on Climate Change's Fourth Assessment Report (AR4) in 2007 triggered this heightened awareness, prompting a wave of new research (Pachauri and Reisinger 2007). Researchers began to employ more comprehensive empirical analyses, employing Geographic Information Systems (GIS) for spatial analysis of migration patterns and environmental changes (Barassi et al. 2018; Mijani et al. 2022). Mixed methods research, which combined qualitative and quantitative approaches, provided a richer understanding of climate migration (Hoffmann et al. 2021). Thematic areas of focus expanded, including rural–urban migration, impacts of climate change on agricultural practices, and the role of migration as an adaptation strategy (Patnaik and Narayanan 2015; Jha et al. 2018). These new areas of study reflected a broader recognition of the diverse ways climate change could influence human movement and livelihoods. From 2015 to 2022, climate migration research experienced exponential growth (Liang et al. 2023), reflecting the growing urgency and importance of the issue in global climate policy discussions. The 2015 Paris Agreement underscored the need for understanding and addressing climate-related migration, as it aimed to limit global warming and mitigate its impacts, including those on human populations (Lu et al. 2016; Silchenko and Murray 2023). Advanced analytical techniques emerged, providing deeper insights into the long-term impacts of climate change on migration patterns. Big data analytics allowed researchers to analyze large datasets, uncovering trends and patterns that were previously difficult to detect. Longitudinal studies offered valuable information about the sustained effects of climate change on migration decisions and outcomes. Key themes during this period included vulnerability, adaptation strategies, and socio-economic impacts of climate migration on both sending and receiving communities.

The bibliometric method was selected for the present study and put into practice (Hualin et al. 2020). Bibliometric is a potent analytical instrument that is well known for providing an organized and trustworthy examination of large amounts of data (Fu and Waltman 2022). Through the use of bibliometric analysis, researchers hope to identify the broad themes, patterns, and connections found in the body of current research literature (Barassi et al. 2018). The method in question enables a systematic analysis that provides a thorough synopsis of the current body of research, as emphasized by Aria and Cuccurullo (2017). Essentially, the researchers aimed to obtain a better understanding of the larger research landscape in their topic of interest by choosing the bibliometric technique (Wang et al. 2018). They were able to gain a deeper understanding of the subject matter by using this strategy, which helped them to discover influential works, uncover developing areas of emphasis, and synthesize important patterns (Liu et al. 2019).

2.1.1 Insights into Knowledge Growth via Bibliometric Techniques

The bibliometric study has gained popularity due to its unique properties, which synthesize current information and analyze research trends in a given field over time (Moscoso et al. 2018; Priovashini and Mallick 2021). Defining keywords is a crucial step in conducting bibliometric analysis, as it sets the foundation for the search process and determines the scope of the study (Ellegaard 2018; Maretti et al. 2019). Key concepts include synonyms, related terms, Boolean operators, truncation, and wildcards. A balance between specificity and sensitivity is essential, with broader keywords yielding more results but potentially lower relevance (Huang et al. 2022). Language and terminology should be inclusive to ensure comprehensive coverage. Documentation is essential for transparency and replicability of the bibliometric analysis. Database selection is crucial, with several academic databases available, each with its strengths, weaknesses, and disciplinary focus (Aria and Cuccurullo 2017). Common databases include Web of Science, Scopus, and Google Scholar. The search strategy in bibliometric analysis involves defining research objectives, identifying keywords, selecting search filters, incorporating controlled vocabulary, adjusting search strings, exploring citation networks, documenting the search strategy, and managing search results (Wei et al. 2015; Becerra Ruiz et al. 2020). Visualization in bibliometric analysis uses graphical representations to communicate patterns, trends, and insights from scholarly literature (Huang et al. 2022).

Khavarian-Garmsir et al. (2023) identified four primary thematic clusters in climate migration research: water-related hazards, forced migration, climate change and adaptation, and population studies and policymaking (Maretti et al. 2019). These clusters address distinct aspects of how climate change influences human migration patterns (Mallick et al. 2021). Water-related hazards including floods, droughts, and sea level rise, which have a direct influence on human settlements and livelihoods and frequently cause migration, are becoming more and more useful for bibliometric study (Hou et al. 2015; Kim et al. 2021). These hazards undermine agricultural productivity, damage infrastructure, and pose immediate threats to safety (Guo et al. 2019). Forced migration focuses on the displacement of populations with low adaptive capacity, emphasizing the role of environmental changes in exacerbating conflicts and resource scarcity. This cluster explores the intersections between environmental changes and socio-political factors, such as resource scarcity exacerbated by climate change (Ellegaard 2018). Intersection of migration policies with broader climate change and environmental policies is another issue of research. Further bibliometric research focuses on the socio-economic impacts of climate migration, the role of policy in facilitating or hindering migration, and the development of frameworks to manage climate-induced migration sustainably (Wang et al. 2018; Mallick et al. 2021). This includes evaluating existing policies, identifying gaps, and proposing new strategies to address the complexities of climate migration. Despite significant advancements, several challenges remain in the field of climate migration research. One major challenge is monitoring migration and climatic events, which requires integrating diverse data sources and developing methodologies to establish causal links between climate change and migration (Becerra Ruiz et al. 2020). Additionally, there is a need to investigate contextual factors and mechanisms that influence migration decisions, such as socio-economic conditions, cultural norms, and local governance structures (Shukla et al. 2019). Another challenge is assessing the socio-economic impacts of climate migration on affected communities, considering both the sending and receiving communities, examining how migration affects social cohesion, economic stability, and resource distribution (Maretti et al. 2019). Overall, addressing these challenges will require methodological innovation and interdisciplinary collaboration. By integrating insights from environmental science, social science, economics, and policy studies, researchers can develop more robust frameworks for understanding and managing climate-induced migration (Rana 2020; Islam et al. 2022).

Future scholarly endeavors should be oriented toward confronting these challenges and delving into overlooked domains, such as the ramifications of climate-induced migration on social cohesion and the dynamics of sedentary populations, particularly those ensnared by economic or social constraints. The amalgamation of interdisciplinary methodologies and the harnessing of emerging data reservoirs, such as those gleaned from social media and mobile communication platforms, hold promise in furnishing a more exhaustive comprehension of the complexities inherent in migration catalyzed by climate shifts. By augmenting our insight into the mechanisms and repercussions of climate-induced migration, academia stands poised to furnish more efficacious policies and frameworks aimed at the regulation and alleviation of its impact. Subsequently, the ensuing section elucidates the methodology, delineating its application and the generation of requisite data for analysis. This is followed by an explication of findings in five distinct subsections, encompassing temporal, source, spatial, and lexical network analyses, alongside the thematic evolution of climate migration society policy (CMSP). Ultimately, we engage in a discussion of these findings and draw conclusions, while also acknowledging the limitations intrinsic to this study.

The variability in the availability of full-text articles across different databases could have impacted the thoroughness of the literature review. Since not all databases contain the same range of articles, the pool of sources for review may have been limited, potentially leading to gaps in coverage. This variance in availability might result in an incomplete understanding of the research landscape on climatic migration. Even though thematic coding was carried out systematically to identify recurring themes and topics within the literature, qualitative analysis inherently involves a degree of subjectivity. Different researchers may interpret the same text differently, leading to potential bias in the identification and classification of themes. To address this issue, efforts were made to enhance the reliability of the analysis through inter-coder reliability checks. These checks involved multiple researchers independently coding a subset of the data and then comparing their results to ensure consistency and reduce the impact of individual biases. By employing such measures, the study aimed to increase the validity and rigor of the qualitative analysis, thereby enhancing the credibility of the findings. Since bibliometric analysis is an iterative process, new research is always being released. To take into account fresh advancements and trends in the literature, think about revising your analysis from time to time.

Novelty of the work is that the interpretation or analysis was done concerning with India and Bangladesh as both countries are facing the adverse impact of climatic change. Cyclonic hazards in this area expose coastal residents, mostly those on the Sundarbans, and push those who are less able to withstand the negative consequences of climate hazards to migrate outside. The primary focus of this study was on areas that are more pertinent to climate migration but have received less attention up to this point, thus it should receive more attention eventually.

2.2 Materials and Methodology

2.2.1 Research Design

Empirical research on climate migration consists of macro-studies assessing regional or national migration, and micro-studies focusing on individual and household movements. Various data sources are used, including country-level data and specialized case studies. Meta-analyses show that results may vary depending on the analytical level used, with projected migratory patterns varying across micro- and macro-level studies. Here, in this research on climatic migration from 1970 to 2024 used a mixed methods approach, combining qualitative and quantitative methods, for an in-depth analysis of trends, dimensions, and methodological advancements.

2.2.2 Data Collection

The study involved a comprehensive review of literature sourced from major academic databases like Scopus and Web of Science. Keywords utilized in the search encompassed terms such as "climatic migration," "climate change and migration," "environmental migration," and "migration adaptation strategies." Publications spanning from 1970 to 2023 were considered to offer a historical context and capture recent developments (Fig. 2.1). Articles were included based on their relevance to climatic migration, peer-reviewed status, and availability of full text. For bibliometric analysis, VOSviewer and R Studio were employed to visualize the co-occurrence of keywords, authorship, and contributions by country. Metadata from selected publications, including titles, authors, publication years, and abstracts, were extracted for analysis. Additionally, SciMAT software facilitated qualitative analysis. The study monitored the number of publications per year to discern trends over time, especially in connection with significant events like the release of IPCC reports. Thematic analysis was carried out to categorize the research's focus areas, such as theoretical frameworks, empirical studies, and policy analysis. Research methodologies were examined decade by decade to identify shifts in approaches and advancements in techniques. Methods were classified into theoretical frameworks, case studies, surveys, quantitative analyses, GIS, meta-analyses, mixed methods, longitudinal studies, and big data analysis.

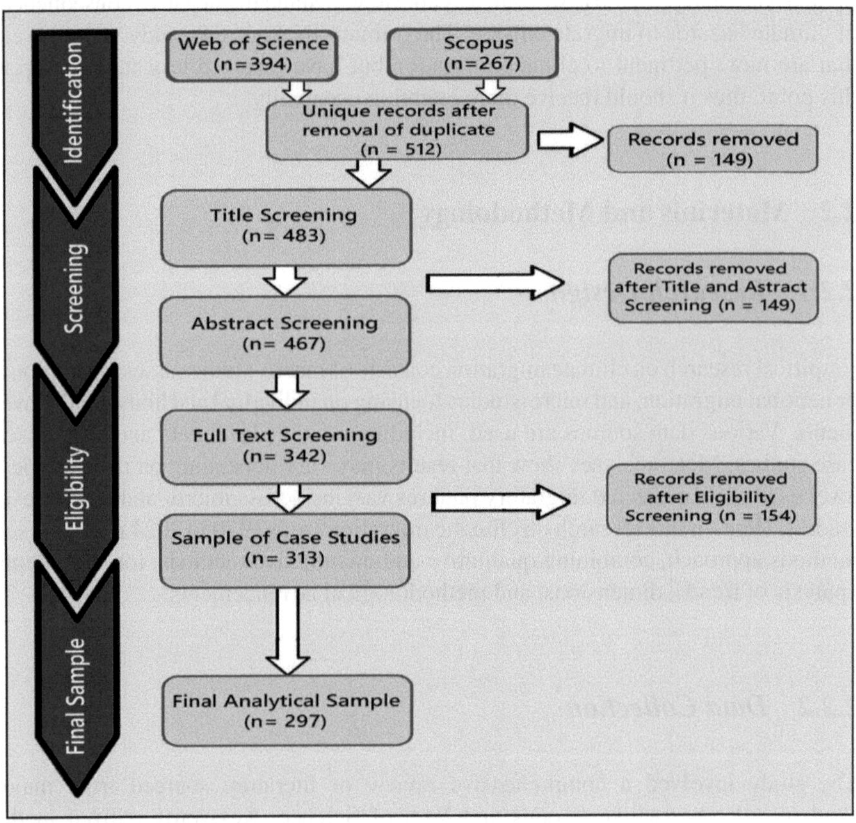

Fig. 2.1 Methodological flowchart of the bibliometric study

2.2.3 *Data Analysis*

Graphs were employed to illustrate the publication trends over time and the influence of IPCC reports on research activity. Summary statistics were computed to offer insights into research output across decades, countries, and methodologies. Qualitative content analysis was utilized to uncover recurring themes and topics in the literature, achieved through coding abstracts and full texts of articles. These findings were synthesized into a narrative detailing the progression of climatic migration research, emphasizing significant milestones and shifts in focus. VOSviewer facilitated the creation of network diagrams, displaying keyword co-occurrence and relationships between research topics. Additionally, it mapped international collaboration networks by examining co-authorship and citation patterns, enabling the identification of thematic groups and collaborative patterns within the research field.

2.3 Results

2.3.1 *Trends of Research Work on Climate Migration*

From 1990 to 2020, there has been a surge in publications on climatic migration, highlighting the growing importance of understanding the relationship between climate change and human migration. The severity and frequency of climate-related events, such as hurricanes, cyclones, droughts, and heat waves, have prompted researchers, policymakers, and communities to delve deeper into the complex relationship between climate change and migration. The release of key IPCC Assessment Reports, such as AR4 and AR5, has shaped the discourse on climatic migration, providing comprehensive assessments of scientific evidence, impacts, and adaptation strategies. Interdisciplinary collaboration and methodological advancements have fueled the proliferation of publications on climatic migration, enhancing our understanding of its socio-economic, political, and environmental implications.

2.3.1.1 Quantification of Research Associated with Climate Migration

The number of publications on climatic migration was relatively low and stable during the period discussed, suggesting that academic interest in the topic was in its early stages. However, around 2007, there was a noticeable increase in publications due to the release of the IPCC Fourth Assessment Report (AR4), which drew attention to the impacts of climate change on migration patterns. The AR5 report, released in 2014, provided updated insights into climate change impacts, emphasizing the link between climate change and migration. This report significantly impacted the research community, leading to a surge in related research activities and publications. Following 2014, there was a continuous upward trend in publications, reaching a peak in 2020, indicating a growing and sustained interest in climatic migration research. Factors contributing to this include increasing global awareness, enhanced support for climate-related research from governments and international organizations, and improvements in data collection and analysis techniques. The IPCC reports (AR4 and AR5) served as critical inflection points, correlating with significant increases in publications and shaping research agendas and priorities. They acted as catalysts for research activity by providing authoritative assessments of climate change impacts and highlighting the urgency and importance of studying climatic migration.

Overall, the graph illustrates the significant evolution of climatic migration research over the past three decades, with the release of key IPCC reports playing a pivotal role in driving research activity and increasing awareness of the importance of understanding and addressing the impacts of climate change on migration. Figure 2.2 displays a significant growth in the literature during 2014–2023; at the very least, the trend line itself supported this claim. Additionally, it is evident from Fig. 2.2 is the substantial increase in the frequency of published papers that occurred after 2014. In conclusion, the increasing number of publications on climatic migration from 1990 to 2023 reflects the growing recognition of this issue as a critical dimension of climate change. The release of key IPCC Assessment Reports, coupled with interdisciplinary collaboration and methodological advancements, has propelled research and policy engagement in this field. However, addressing the challenges posed by climatic migration requires concerted efforts across multiple fronts, including scientific research, policy formulation, and international cooperation, to mitigate risks, protect vulnerable populations, and build resilience in the face of environmental change.

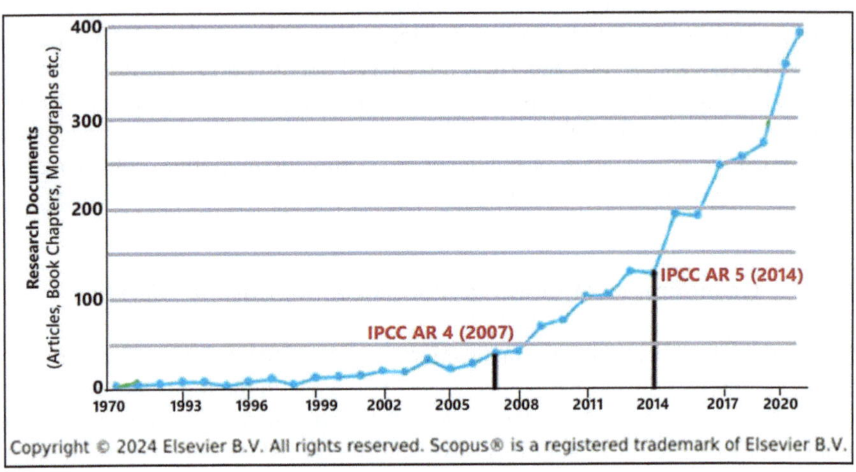

Fig. 2.2 Publication trend on cyclone-induced vulnerability and migration research from 1970 to 2023

2.3.1.2 Thematic Evolution of Climate Migration Research

The trend of climatic migration literature from 1970 to 2024 shows a significant increase in research activities over the decades, with notable shifts in focus areas and methodological advancements. Initially dominated by theoretical frameworks, the research has evolved to include comprehensive empirical analyses, quantitative methodologies, and systematic reviews, reflecting the growing complexity and importance of understanding climate-induced migration. The trajectory of literature on climatic migration from 1970 to 2023 reveals a notable surge in research activities over the decades, accompanied by discernible shifts in focus areas and methodological advancements (Table 2.1). Initially characterized by theoretical frameworks, research has evolved to encompass comprehensive empirical analyses, quantitative methodologies, and systematic reviews, reflecting the increasing complexity and significance of understanding climate-induced migration. In the early stages, research was nascent regarding the correlation between climate and migration. Scholars primarily concentrated on formulating initial theoretical models and frameworks to conceptualize how environmental changes could impact human migration. This foundational work laid the groundwork for subsequent empirical studies. As time progressed, researchers began empirically documenting migration patterns using qualitative methods such as case studies and interviews. These methods provided rich narratives and real-world examples of how communities responded to climatic changes, marking a transition from purely theoretical work to the collection of empirical evidence.

The subsequent decade witnessed the growing prominence of surveys as researchers sought to systematically gather data from larger populations. This facilitated a more detailed empirical analysis of various migration drivers, including

Table 2.1 Thematic evolution of literatures on climate migration during 1970–2023

Decade	Primary methods	Key focus	Notable trends/changes
1970–1979	Theoretical frameworks	Conceptualizing climate impacts on migration	Emergence of basic models and theoretical approaches
1980–1989	Case studies, interviews	Empirical documentation of migration patterns	Introduction of qualitative case studies and interviews
1990–1999	Survey-based research	Empirical analysis of migration drivers	Increased use of surveys for primary data collection
2000–2009	Quantitative analyses, GIS	Mapping migration flows, climatic impacts	Adoption of GIS for spatial analysis, focus on quantitative data
2010–2019	Meta-analyses, mixed methods	Systematic reviews, integrating multiple data	Rise of mixed methods, comprehensive meta-analyses
2020–2023	Longitudinal studies, big data	Long-term impacts, digital trace data	Use of big data and longitudinal studies for deeper insights

Source Developed by the researcher

economic, social, and climatic factors. Surveys yielded quantitative data that could be analyzed to identify patterns and correlations. Technological advancements, notably the adoption of Geographic Information Systems (GIS), became significant during this period (Table 2.1). GIS enabled researchers to map and analyze spatial data, offering visual representations of migration flows and their association with climatic changes. This facilitated more precise analysis of data and enhanced understanding of migration dynamics. Quantitative methods continued to evolve, with researchers conducting meta-analyses to systematically review and synthesize findings from multiple studies. This approach helped identify broader trends and generalize conclusions across different contexts. The emergence of mixed methods research, combining qualitative and quantitative approaches, provided a more comprehensive understanding of the complexities surrounding climate migration. In recent years, the utilization of longitudinal studies has become more prevalent, offering insights into the long-term impacts of climate change on migration decisions. Additionally, the advent of big data, including digital trace data from sources like social media and mobile phones, has enabled researchers to analyze large datasets and uncover patterns that were previously challenging to discern, offering a more dynamic and real-time perspective on migration dynamics.

The methodology for studying climatic migration has evolved from theoretical frameworks to advanced empirical and analytical techniques. Each decade brought significant advancements in methods, reflecting growing complexity and depth of research in this field. Recent trends emphasize the integration of multiple data sources, advanced analytics, and long-term studies to capture the comprehensive impacts of climate change on migration patterns. This trend analysis and the corresponding diagram highlight the progression and diversification of research methodologies over time, showcasing how the field has adapted to address the growing challenges and complexities associated with climatic migration (Table 2.1).

2.3.1.3 Country-Wise and Author-Wise Research on Climate Migration

The map shows the global scientific production, where dark color indicating higher outputs. In terms of climate migration research, the leading countries are the U.S. and Australia, while India and Bangladesh show moderate contributions (light blue for India, with Bangladesh likely in a similar range) despite being highly vulnerable to cyclonic hazards, particularly in the Sundarbans region (Fig. 2.3). India and Bangladesh are with growing research infrastructure, and for it the contribution is also increasing to climate migration studies, especially focused on internal displacement and coastal vulnerabilities, such as in the Sundarbans. From the network map, it is clear that both countries have not significant collaborative work on this filed though both country faces same climatic vulnerability and more important both countries share the Border line, highlighting the need for greater collaborative research efforts to address the shared vulnerabilities in the Sundarbans.

The volume of chosen literature on the highlighted theme—climate migration—confirms the subject's significance in a world that is changing. In this case, the selected

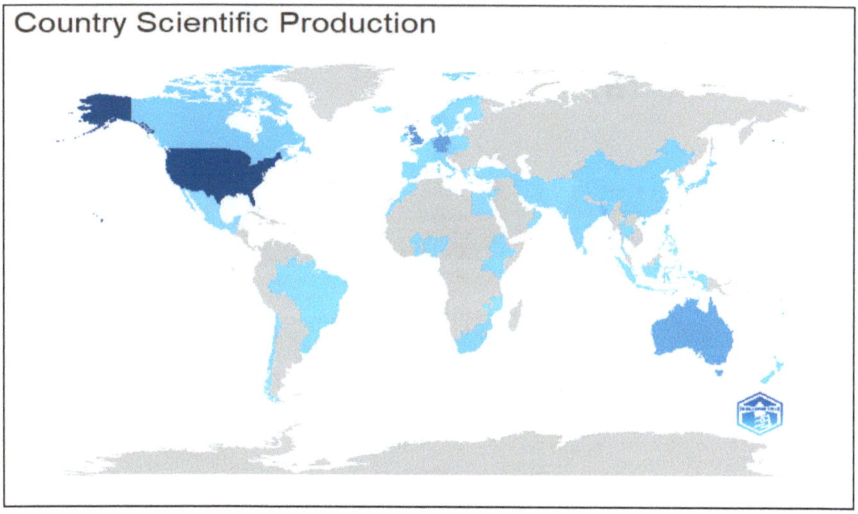

Fig. 2.3 Country scientific production

research papers for bibliometric analysis cover the years 1970–2023 or roughly the last 50 years. The 297 publications that were obtained were from the Web of Science (WoS) database, and their location was indicated in Fig. 2.4a. The diagram (Fig. 2.4a) depicts the global research landscape on climatic migration, revealing the major countries leading the field and their collaborative networks. The United States is the central hub, with the largest node representing its significant contribution to the field. China, the United Kingdom, and France are also prominent contributors, demonstrating robust international collaborations and citations. Brazil, Denmark, India, and Germany are active participants in the field, demonstrating significant engagement and citations. The presence of different colored clusters signifies regional or thematic collaborations within climatic migration research, suggesting that countries within the same region or with similar research focuses often collaborate closely. Countries at the periphery, such as Malaysia, Bangladesh, Indonesia, and Vietnam, have smaller nodes with fewer connections, indicating lower research output and fewer collaborations or citations compared to central countries. However, their presence in the network highlights their engagement in climatic migration research, albeit to a lesser extent. The diagram offers a comprehensive overview of the global collaboration pattern in climatic migration research, illustrating the interconnected nature of research efforts. Major hubs like the United States, China, and the United Kingdom facilitate widespread international cooperation, contributing to advancing our understanding of climatic migration and developing effective strategies to address its complex challenges.

In this section, top ten country-wise total publications on climate migration represent through a bar diagram. Bar diagram gives a visual outline of data enabling the researchers to know about average values of the given dataset. Here to show the

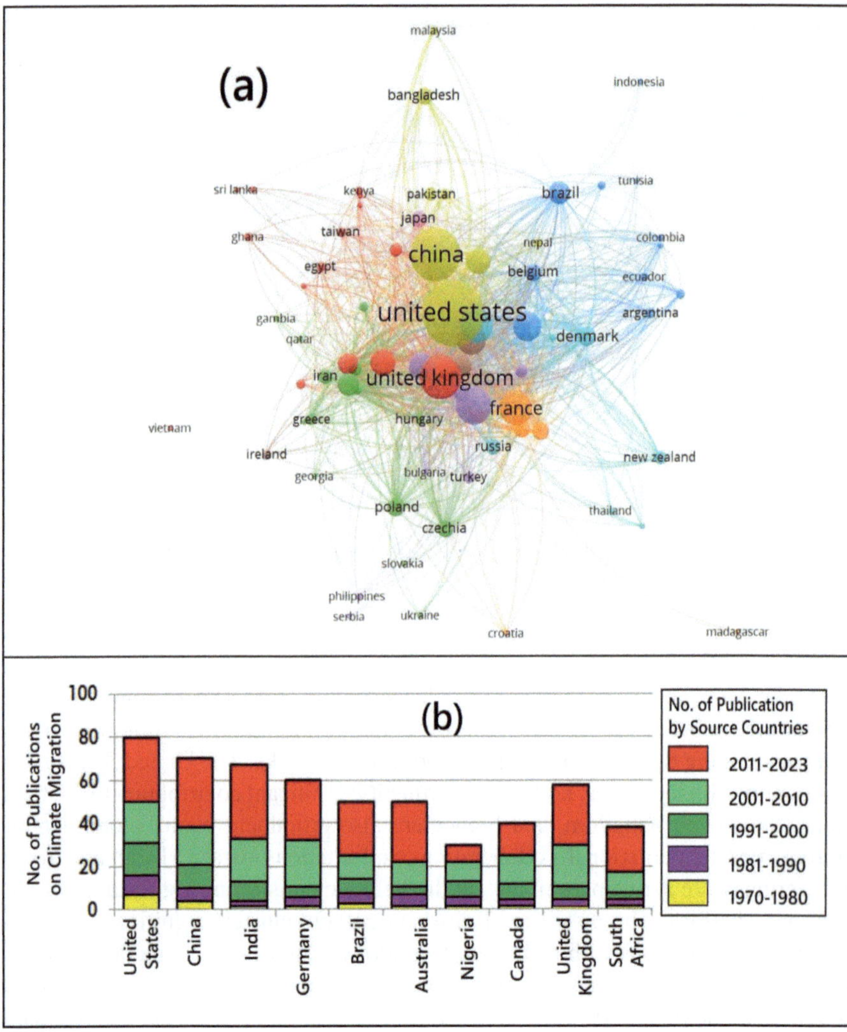

Fig. 2.4 **a** Countries associated with climate migration research from the retrieved publications (1970–2023) and **b** top ten countries contributing to the field of climate migration research (1970–2023)

variability and array of research publications in the mentioned research fields, the top ten productive countries have been chosen for cartographical layout through a bar diagram (Fig. 2.4b). The overall impression of the bar diagram glimpses that the average rate of publications for all the counties is not uniform. Figure 2.4b also demonstrates the country-wise degree of dispersion in the published literature among the topmost ten productive countries during 1970–2023. Overall, we observe a changing landscape in climate migration research, as established research powerhouses such as the United States and Germany are rapidly being surpassed

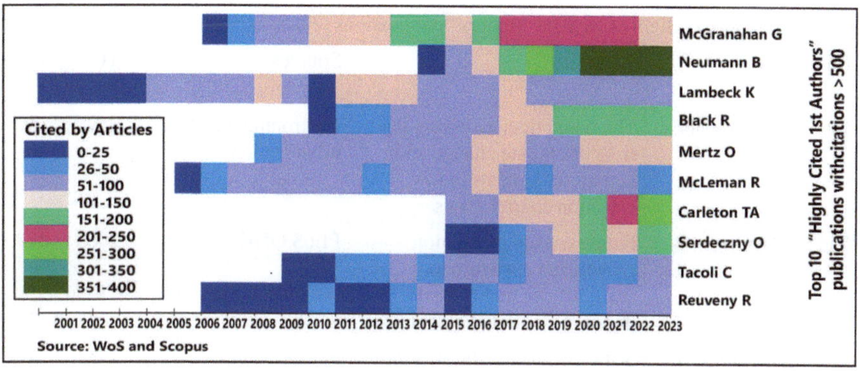

Fig. 2.5 Top 10 authors with yearly publication frequency in the field of climate migration from 2001 to 2023

by growing nations such as China, India, and Brazil. This change emphasizes how dynamic international research is and how crucial innovation and adaptation are to preserving research competitiveness.

In terms of total citations, McGranahan leads with 2,578 citations, followed by Neumann with 2,063 citations. Neumann's article on climate migration, published in PLoS ONE in 2016 (Fig. 2.5), has the highest citation rate, averaging over 200 citations per year, highlighting his significant contribution over the last decade. Table 2.2 also lists one publication by Lambeck (TS = 1,784) and two by Black (TS = 1,352) among the top ten most influential contributions to climate migration research based on citations. Notably, Lambeck is the longest active researcher in the dataset, with publications spanning from 2001 to 2023 (Fig. 2.5). Black is the fourth leading contributor, along with Mertz, in terms of total citations (TS = 1,320), followed by McLeman (TS = 1,190) and Carleton (TS = 1,160), each with one highly cited publication. Serdeczny ranks eighth with 768 total citations.

Analyzing these scholarly contributions highlights their influence on the development of this research field. Identifying the most cited and an influential article helps understand the knowledge structure in this area (Bamel et al. 2020). Citation analysis is a valuable tool for assessing the quality of research contributors (Guo et al. 2019; Khan and Ho 2012). According to Biswas et al. (2021) and Khan and Ho (2012), publications cited more than 100 times are typically considered "highly cited" or "classic." Table 2.2 lists the top ten influential publications that meet this threshold.

2.3.1.4 Subject Categories of the Research

Following the study's objective, the retrieved publications were organized into 47 subject categories in the Web of Science database. We focused on quantitative empirical work examining the initial stage of climate migration, specifically the relationship between climatic impacts, vulnerability, risk, and migration. This focus influences the

Table 2.2 Top 10 "Highly Cited 1st Authors" publications (citations > 500)

Rank	1st author	Title	Sources	TC	TC Y^{-1}
1st	McGranahan G	The rising tide: assessing the risks of climate change and human settlements in low elevation coastal zones	Environment and urbanization	2578	152
2nd	Neumann B	Future coastal population growth and exposure to sea-level rise and coastal flooding—a global assessment	PLoS ONE	2063	229
3rd	Lambeck K	Sea level change through the last glacial cycle	Science	1784	78
4th	Black R	The effect of environmental change on human migration	Global environmental change	1352	104
5th	Mertz O	Farmers' perceptions of climate change and agricultural adaptation strategies in rural Sahel	Environmental management	1320	88
6th	McLeman R	Migration as an adaptation to climate change	Climatic change	1190	66
7th	Carleton TA	Social and economic impacts of climate	Science	1160	145
8th	Serdeczny O	Climate change impacts in Sub-Saharan Africa: from physical changes to their social repercussions	Regional environmental change	768	96
9th	Tacoli C	Crisis or adaptation? migration and climate change in a context of high mobility	Environment and urbanization	629	45
10th	Reuveny R	Climate change-induced migration and violent conflict	Political geography	616	583

disciplinary representation of authors. Table 2.3 lists the top ten subject categories in the field of climate change and agricultural vulnerability scenarios. The results show that the published documents were classified into various subject areas, highlighting the multidisciplinary nature of the research theme. It is clear that Environmental Science (38.72%) is the leading subject category, followed by Disaster Management (15.82%), and Meteorology and Atmospheric Sciences (13.80%). In the arts, humanities, and social sciences, Geography (12.12%) and Economics (7.41%) are the most prevalent subject categories. The analysis indicates that research on climate change and agricultural vulnerability has gradually expanded into different subject categories, although environmental science remains a dominant field (Fig. 2.6).

Table 2.3 Top 10 subject categories of the retrieved publications

WoS subject categories	Total publication	% share	WoS subject categories	Total publication	% share
Environmental sciences	115	38.72	Population studies	14	4.71
Disaster management	47	15.82	Sociology	9	3.03
Meteorology and atmospheric Sciences	41	13.80	Agricultural science, multidisciplinary	5	1.68
Geography	36	12.12	Development studies	5	1.68
Economics	22	7.41	Geosciences, multidisciplinary	3	1.01

Source Developed by the researcher

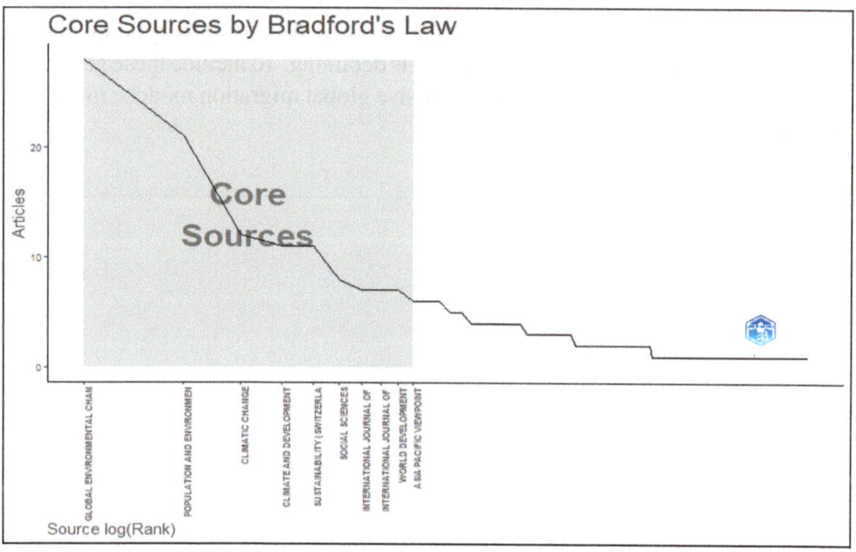

Fig. 2.6 Core sources at global level of the research

2.3.1.5 Thematic Evolution

The diagram (Fig. 2.8) depicts that the emphasis on refugees and international migration indicates how important these topics are to conversations about migration, particularly in light of climate change and natural disasters. These topics are well established. Scholars often study human migration, frequently making connections between it and concerns about human rights, adaptation tactics, and policy development. These themes' inclusion in this quadrant highlights both their vital significance and the vigorous, ongoing research being done in this area. Although they are less developed than driving themes, climate change, migration, and adaptive management appear to be important topics in the field of study, as indicated by their placement in this quadrant. The mechanisms underlying climate-induced migration and adaptive management are important and continuing concerns; nevertheless, further research may be required to completely understand these linkages. To comprehend the wider effects of environmental risks on human migration, these topics are crucial. This quadrant identifies fields with substantial, focused research but little overall influence (Fig. 2.7). Research on the Khulna and Sub-Saharan regions points to extensive localized studies on migration and vulnerability brought on by climate change, but it's possible that these results haven't been widely incorporated into the larger conversation yet. Although these issues have been extensively researched in various contexts, it's possible that they haven't yet had an impact on conversations about global migration. The fact that places like the Carteret Islands and Papua New Guinea feature in this quadrant indicates that research in these locations is either beginning to grow or is already declining. These places are among the most susceptible to relocation brought on by climate change, especially as a result of rising sea levels, but their inclusion here suggests that either fresh case study is being made of them or that the importance of their research is declining. To include these geographically isolated examples in more comprehensive global migration models, more work could be required.

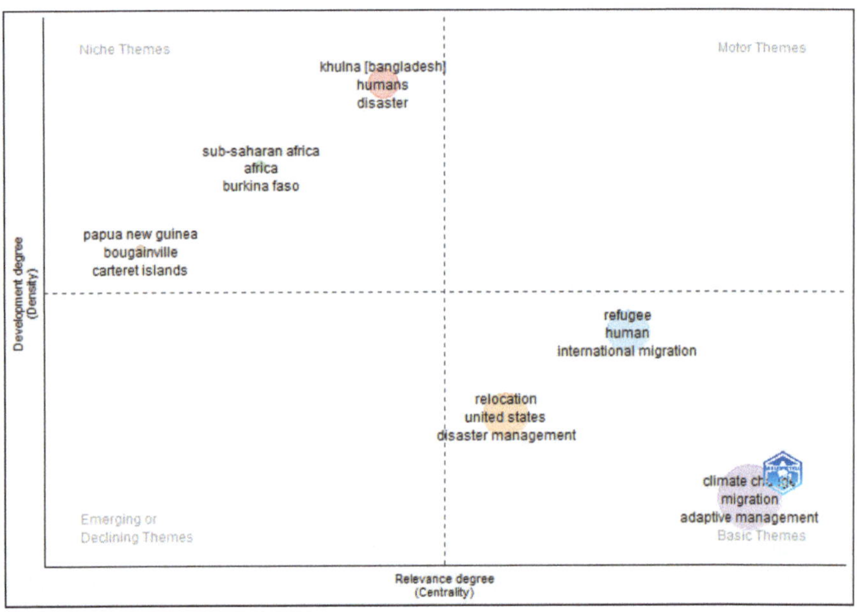

Fig. 2.7 Thematic evolution

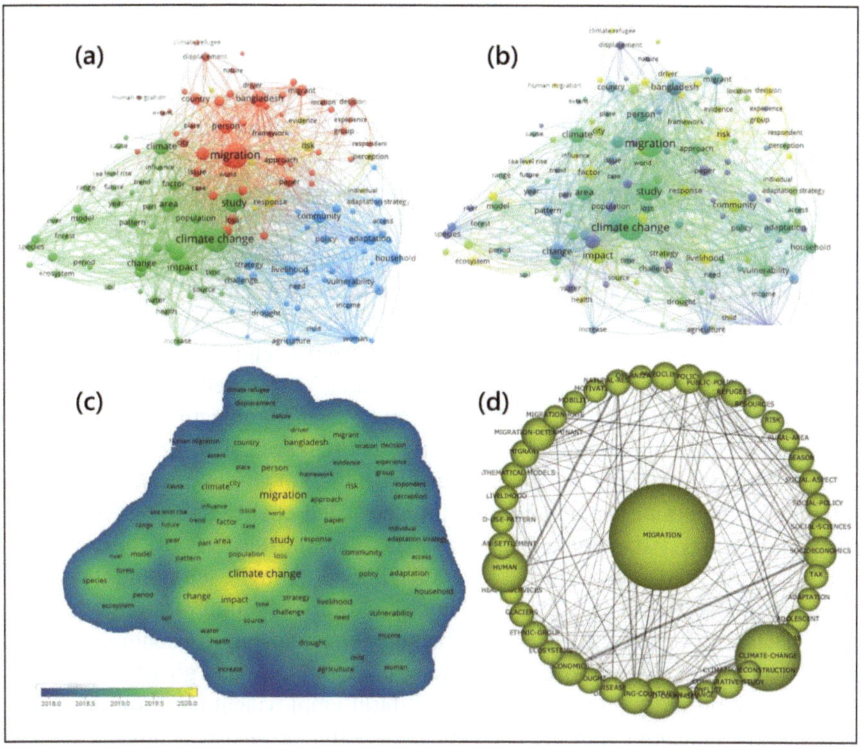

Fig. 2.8 Nature of visualization—**a** network, **b** overlay **c** density, and **d** connectivity analysis

2.4 Keyword Visualization on Climate Migration

Keyword visualization is a technique used in bibliometric surveys to identify and display key terms within publications, uncovering trends, patterns, and relationships within a research field, enhancing the accessibility and visual appeal of complex data. VOSviewer is a tool for visualizing and analyzing bibliometric data. It involves selecting a relevant database, defining search terms, and exporting the search results in a format compatible with VOSviewer. Data preparation involves cleaning and standardizing the data, ensuring it is properly formatted for VOSviewer. The data is then loaded into VOSviewer, which creates a map and analyzes the data. The visualization settings are adjusted to better interpret the data. The data is then examined to identify key themes, clusters, trends, and trends. The findings are documented in insights, such as major research themes, influential papers, key authors, and the overall structure of the research field. The visualizations are then exported in high-resolution formats for inclusion in a report or presentation. The report includes detailed descriptions of the findings and visualizations. VOSviewer offers a comprehensive and visually engaging way to analyze and present research trends, relationships, and thematic structures in a given field.

2.4.1 Nature of Visualization (Network, Overlay, and Density)

The network diagram illustrates the thematic connections and relationships among various research topics in the field of climate change and migration. Each node (representing a keyword or research topic) varies in size based on its frequency or importance in the research field. Larger nodes indicate more frequently occurring or significant topics. Different colors represent distinct clusters or groups of related topics, each color signifying a unique thematic group within the research field. The lines connecting the nodes represent co-occurrence or thematic relationships between topics, with thicker and more numerous lines indicating stronger and more frequent connections. "Climate Change" is the largest and most central node, indicating it is the primary focus of the research field. "Migration," influenced by climate change, is another central node closely linked to climate change, reflecting its significance in the context of climatic impacts. Migration is often seen as a coping strategy for climate change, making "Adaptive Management" another significant node, highlighting its importance in addressing climate-related challenges. The Red Cluster focuses on policy, risk management, and socio-economic aspects of climate change and migration. The Green Cluster covers livelihood, drought, agriculture, and rural population. The Blue Cluster examines household survey, coping strategies, and crop yield. The Yellow Cluster focuses on sustainability, urbanization, economic development, and governance approach, addressing urban impacts and governance. Topics like gross domestic product, flood, and off-farm employment are positioned more peripherally, indicating they are less central but still relevant to the broader research landscape. The diagram highlights the interdisciplinary nature of research in climate change and migration, with strong interconnections between different thematic areas. It shows that research on climate change and migration encompasses a wide range of topics, including socio-economic, policy, environmental, and agricultural aspects (Fig. 2.8a).

The color overlay in the Overlay Visualization offers valuable insights into the evolution of research themes over time, highlighting recent trends and emerging areas of focus. This visualization effectively demonstrates the multidisciplinary nature of climate migration research, encompassing human impacts, environmental science, and policy responses. By incorporating a temporal dimension, the visualization helps identify which areas are currently receiving increased attention, indicating shifts in research priorities and emerging trends. Both "migration" and "climate change" are central and well-connected nodes in the visualization, underscoring their pivotal roles in the research field. The yellow shading around these terms suggests a recent surge in research focus on these topics. Additionally, the terms "adaptation," "community," and "policy" show significant interconnections and a notable increase in recent research interest. This indicates a growing emphasis on understanding and improving adaptation strategies in response to climate impacts. The keyword "Bangladesh" is prominently featured and shaded yellow, signifying ongoing and recent research interest in this region. This focus is likely due to Bangladesh's high vulnerability to climate impacts and significant migration patterns, making it a critical area of study within the broader context of climate migration research (Fig. 2.8b).

This density visualization generated by VOSviewer using Scopus bibliometric data provides a heat map representation of keyword frequency and density within the literature on climatic migration. The density visualization helps identify the most heavily researched areas and the distribution of focus across different topics. The color gradient from blue to yellow indicates the density of keyword occurrences. Yellow areas represent higher keyword density (more frequent research focus), while blue areas represent lower density (less frequent research focus). The keywords are spread across the map, with their density indicating the intensity of research activity associated with each term. The density of keyword has been categorized into three classes (Fig. 2.8c).

The diagram shows that "migration" is the primary focus in the literature on climatic migration, with a high frequency of occurrence. Climate change is also a core research area, with a strong connection to various topics related to climatic migration. Bangladesh is a significant regional focus due to its vulnerability to climate change and associated migration patterns. Research on the impacts of climate change, including how it drives migration, is significant. Adaptation strategies are emphasized, emphasizing the importance of policy and community responses. The green areas highlight community-level responses and resilience to climatic changes, policy measures related to climatic migration and adaptation, and risks associated with climatic migration. Household-level impacts and adaptation strategies are also highlighted. Environmental factors and terms like species, river, forest, soil, and ecosystem are less frequent research compared to broader themes of migration and climate change.

2.4.2 Connectivity Analysis

Migration is the largest node in the network, indicating that "migration" is the most frequently occurring keyword in the dataset. This reflects the central focus of research on climate-induced migration. As the second largest node, "climate change" is closely linked to "migration." This shows a significant amount of research examining the impact of climate change on migration patterns. The "human" node suggests a substantial focus on the human dimension of migration, including aspects such as human behavior, human rights, and the impacts of migration on people due to climate change (Fig. 2.8d). This keyword highlights studies on the resilience of communities and individuals facing climate-induced migration. Research in this area often examines how people and communities adapt to changing environmental conditions.

The presence of keywords related to adaptation strategies indicates that a significant portion of the research focuses on how migration serves as an adaptation strategy to cope with climate impacts. Nodes like "policy" and "public policy" suggest a strong interest in the role of policy responses in managing climate-induced migration. This includes creating and implementing public policies to address migration issues. Research also highlights the economic impacts of climate-induced migration, such as the financial costs and benefits to both source and destination areas. The literature reflects research into health-related issues, examining how climate change and migration affect the spread of diseases and the health of migrant populations. Studies indicate research on the potential for conflicts arising from climate-induced migration, including conflicts over resources, social tensions, or geopolitical issues.

The network analysis reveals clusters focusing on environmental factors driving migration, human resilience, and adaptation strategies. Nodes like "policy," "public policy," and "refugee" highlight research on governance, policy responses, and legal aspects of migration. Keywords like "economic," "development," and "livelihood" highlight the economic impacts and development-related aspects of climate-induced migration. This comprehensive overview provides a detailed picture of the research landscape on climate-induced migration, highlighting central themes, significant nodes, and various subthemes identified through network analysis.

2.4.3 Co-citation Analysis

The network diagram (Fig. 2.9) shows the co-citation network in research articles on cyclone migration. The citation frequency of significant articles highlights their academic impact in this research area. Authors like Barnett, Black, and Warner have larger nodes, indicating they are highly cited and influential in the field of climate migration. These authors are central figures in their clusters and have strong co-citation links with other authors. The map is divided into several color-coded clusters, each representing a subfield or specific focus within climate migration research. The red cluster likely focuses on socio-economic impacts and policy aspects of climate migration. The blue cluster might relate to environmental science and the physical impacts of climate change on migration. The green cluster could involve studies on adaptation of strategies and community resilience. Dense areas with many interconnected nodes indicate a well-established and collaborative research community. The presence of several strong co-citation links suggests that certain topics within climate migration are highly integrated and widely studied. Smaller nodes on the periphery of the map represent authors with fewer citations or those who are less central to the main research communities. These authors might be emerging researchers or those who focus on niche topics within the broader field. Adger and Barnett appear to be the most influential, as they have the largest nodes and are central in their clusters. There is significant interconnectivity between different clusters, indicating interdisciplinary collaboration and the multi-faceted nature of climate migration research. Authors like Gemenne, Mueller, and Piguet are also key contributors, as evidenced by their prominent positions and strong co-citation links.

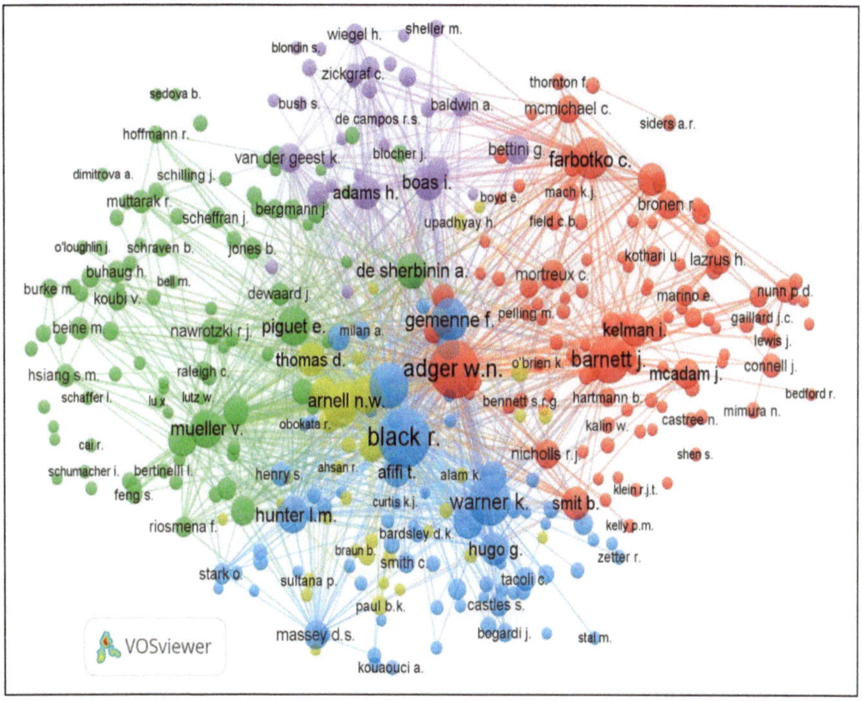

Fig. 2.9 Co-citation analysis of the cited references in the context of cyclone migration since 1970

2.4.4 *Collaborative Work with Countries*

The map illustrates the research collaboration in the area of climate migration with various countries. There have been close ties between Europe and North America, despite the obvious connections between countries like the United States, United Kingdom, and Australia, as well as India and Bangladesh. This indicates that while Bangladesh and India are both very susceptible to migration due to the climate change, their participation in international research networks is comparatively low. The absence of a robust partnership to address the research gap with nations facing comparable climate challenges—particularly in South Asia, Africa, or Latin America—indicates a need for deeper collaborative ties (Fig. 2.10).

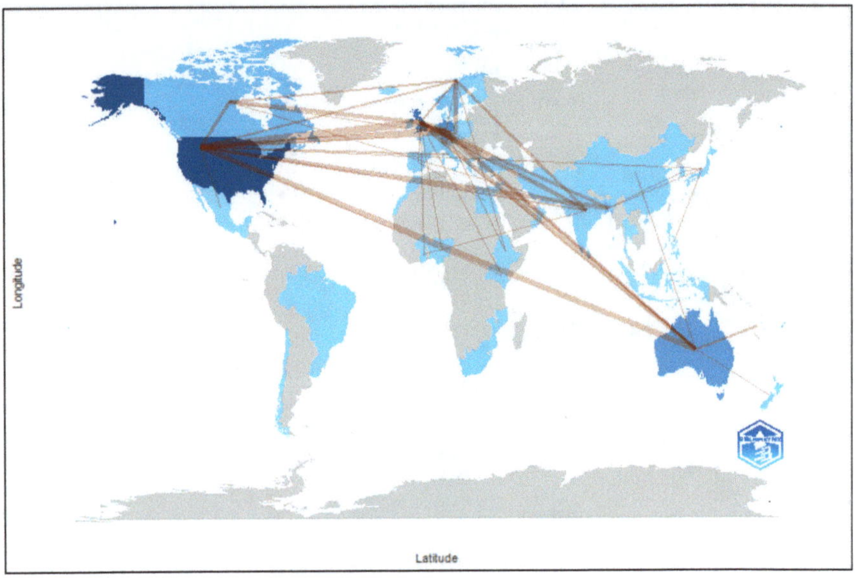

Fig. 2.10 Collaborative work with countries

2.5 Discussions

The bibliometric analysis of climate migration over the last 40 years reveals several significant trends and developments in this field. This section will discuss the primary findings, highlight the conceptual evolution, and identify the gaps and future directions in climate migration research. The analysis of research trends indicates a substantial increase in publications on climate migration from 1990 to 2020. This growth can be segmented into distinct periods, each characterized by shifts in research focus and methodological advancements. The release of the IPCC Fourth Assessment Report (AR4) in 2007 and the Fifth Assessment Report (AR5) in 2014 served as critical inflection points, significantly boosting the number of publications and research activities (Patnaik and Narayanan 2015; Mistri 2019). This trend underscores the influential role of authoritative climate change reports in shaping research agendas and priorities (Sengupta and Samanta 2022; Mijani et al. 2022).

The thematic evolution approach to studying climate migration has evolved considerably. Initially dominated by theoretical frameworks in the 1970s, the research progressed to include case studies and interviews in the 1980s, and survey-based research in the 1990s (Shukla et al. 2019; Liang et al. 2023). The 2000s saw the adoption of quantitative analyses and Geographic Information Systems (GIS), enabling more precise and comprehensive spatial data analysis (Mazumdar et al. 2014; Mallick et al. 2021; Mondal et al. 2022a). The most recent decade has witnessed the rise of meta-analyses, mixed methods, longitudinal studies, and the use of big data. These advancements reflect a growing complexity and depth in understanding climate-induced migration, allowing for more dynamic and real-time perspectives on migration patterns (Lu et al. 2016; Rana 2020; Mondal et al. 2022b). The conceptual evolution of climate migration research has been marked by the development of several key themes. These include water-related hazards, forced migration, climate change and adaptation, and population studies and policymaking (Sarkhel and Mukherjee 2021; Silchenko and Murray; 2023). The analysis revealed that water-related hazards, such as floods, droughts, and sea-level rise, have been a central focus due to their direct impact on migration decisions (Sarkhel and Mukherjee 2021; Mondal et al. 2023). Forced migration has also been a significant theme, emphasizing the involuntary nature of many climate-induced movements (Wang et al. 2018; Islam et al. 2022; Mandal et al. 2022a).

The thematic analysis using VOS viewer identified distinct clusters in the literature. The largest cluster focused on water-related hazards, indicating a strong research interest in how these environmental factors drive migration (Guha and Roy 2016; Kim et al. 2021). Other clusters highlighted the importance of policy responses, socio-economic impacts, and adaptation strategies (Barassi et al. 2018; Becerra Ruiz et al. 2020). These clusters demonstrate the interdisciplinary nature of climate migration research, connecting human impacts, environmental science, and policy responses (Guha and Roy 2016; Islam et al. 2022).

Despite significant advancements, several gaps remain in the literature. One critical gap is the need for more empirical evidence on the mal-adaptation effects of climate migration. While many studies focus on adaptation strategies, less attention has been given to how migration can sometimes increase vulnerability and mal-adaptation to climate change (Ellegaard 2018; Rana 2020; Halder et al. 2021). Future research should explore the characteristics and aspects of migration that negatively affect households' adaptive capacity (Hoffmann et al. 2021; Fu and Waltman 2022). Another gap is the impact of climate migration on social capital. While the effects on physical and financial capital are well documented, the implications for social networks and community cohesion are less understood capacity (Sarkhel and Mukherjee 2021; Fu and Waltman 2022). Additionally, there is a need to clarify whether drought is the most influential environmental factor driving migration or if it is simply more studied due to data availability (Sengupta, and Samanta 2022; Mijani et al. 2022). Future research should also focus on the demographic characteristics and economic factors leading to immobile populations, particularly trapped populations who cannot migrate despite facing severe environmental stressors. Understanding the dynamics of these populations can inform more effective policy interventions (Halder et al. 2021; Mandal et al. 2024).

While "refugees" and "international migration" are central and well developed (motor themes), the foundational work on "climate change," "migration," and "adaptive management" remains in the basic quadrant. This could indicate a gap between theoretical discussions of migration and practical climate adaptation solutions. Researchers may need to focus on integrating adaptive management strategies into broader migration frameworks. The concentration of research on regions like Khulna (Bangladesh) and Sub-Saharan Africa in the niche quadrant may suggest that while localized studies are well developed, they are not as relevant or influential in shaping global migration policies. Strengthening the connection between localized research and global discussions could enhance the relevance of these findings.

2.6 Conclusion

The analysis of research trends on climatic migration from 1990 to 2023 reveals a significant evolution, marked by key milestones and increasing academic interest. Initially, the number of publications was low and stable until 2007, when the release of the IPCC Fourth Assessment Report (AR4) led to a noticeable increase. This trend continued with the introduction of the IPCC Fifth Assessment Report (AR5) in 2014, resulting in a more substantial rise in publications. Post-2014, a steep and continuous upward trend reflects growing global awareness, enhanced support for research, and advancements in data collection and analysis techniques. Over the decades, the focus has shifted from theoretical frameworks to comprehensive empirical analyses and advanced quantitative methodologies. The VOSviewer analysis highlights the United States' central role in research, with significant contributions from China, the United Kingdom, and France, while peripheral countries show less research output. Keyword analysis shows central themes like climate change, migration, and adaptive management, with high research activity in core areas and less focus on specific environmental factors. The systematic analysis reveals central themes such as migration, climate change, human resilience, and adaptation strategies, indicating future research should explore economic and health-related impacts and strengthen the link between policy and human dimensions for effective governance and community resilience.

2.7 Research Gaps and Direction of Future Research

Identifying less connected nodes or smaller clusters in the co-citation network is crucial for recognizing potential research gaps. The absence of a robust partnership to address the research gap with nations facing comparable climate challenges—particularly in South Asia, Africa, or Latin America—indicates a need for deeper collaborative ties. The mechanisms underlying climate-induced migration and adaptive management are important and continuing concerns; nevertheless, further research may be required to completely understand these linkages. Research on the Khulna and Sub-Saharan regions points to extensive localized studies on migration and vulnerability brought on by climate change, but it's possible that these results haven't been widely incorporated into the larger conversation yet. Although these issues have been extensively researched in various contexts, it's possible that they haven't yet had an impact on conversations about global migration.

2.8 Specific Objectives and Goals of the Study

Based on the research gap and direction of future research which have been recognized through bibliometric analysis we have formulated specific objective to establish the title of the research as "cyclone-induced vulnerability and migration." We have already established that Bay of Bengal is most suitable for cyclogenesis and makes the Sundarbans a hotspot region of cyclonic hazards and every single year one to two minimum cyclones hit coastal Sundarbans and destroyed the properties, damaged their assets, and loss the crop production. Coastal residents' livelihoods are therefore in jeopardy, and the severity of the situation has been growing daily, forcing many to leave their homes and work as migrant laborers in order to temporarily support their remaining family members.

The specific objectives of the research are as follows:

Objective 1: to mapping the cyclone-induced potential human migration risk in Pathar Pratima CD Block, Sundarbans, India

To fulfill the objective 1, the following goals have been set:

(a) Evaluate the exposure, sensitivity, and socio-economic adaptive capacity at the village level concerning cyclonic activity in coastal areas affected by cyclones like Bulbul (2019), Amphan (2020), and Yaas (2021), aiming to understand the vulnerability of communities to migration risks.
(b) Analyze the agricultural vulnerability to cyclones and the availability of infrastructure in coastal regions affected by cyclones, focusing on assessing the risk factors that contribute to migration decisions among communities facing cyclonic hazards.
(c) Develop a methodology to quantify integrated vulnerability, including adaptive capacity, and spatially analyze the pattern of migration risk in coastal areas prone to cyclonic activity. This will aim to provide insights into the factors driving migration decisions and potential coping mechanisms for at-risk populations.

Objective 2: to comprehend the nature of migration and the demographic as well as socio-economic condition associated with such migration in the study area (Pathar Pratima CD Block, Sundarbans, India)

To fulfill the objective 2, the following goals have been set:

(a) Investigate cyclones' socio-economic impacts on migrant and non-migrant households in the southern coastal regions of West Bengal, India, aiming to understand how cyclones affect the livelihoods and well-being of coastal populations.
(b) Identify and analyze the motivations underlying rural migration from the Coastal Sundarbans region, considering demographic, social, and economic factors and exposure to cyclones to comprehend the drivers of migration in cyclone-prone areas.

(c) Investigate migration trends related to extreme weather events, particularly cyclones, in the study area. Focus on understanding how demographic, social, and economic aspects influence migration decisions following cyclonic hazards, providing insights for disaster management strategies in the future.

Objective 3: to judge the determinants of migrants, with a specific focus on the impacts of cyclonic events

To fulfill the objective 3, the following goals have been set:

(a) Investigate and identify the influential elements behind cyclone-induced migration by examining variables such as village location, social categories, religion, family structure, gender composition, educational level, and income levels post-cyclone, using logistic regression analysis to understand the relationship between these factors and migration decisions.
(b) Analyze the migration patterns and dynamics by categorizing migration types into cyclonic and non-cyclonic categories and exploring how village location, social categories, religion, family structure, gender composition, educational level, and income levels influence migration decisions, particularly in the context of cyclonic events such as Cyclone Amphan, using logistic regression modeling.
(c) Evaluate the socio-economic impacts of cyclones on migration decisions by examining changes in income levels and workforce participation post-cyclone, assessing how these factors influence migration patterns and decisions within the study area, and providing insights into the resilience and adaptive strategies of affected populations in response to cyclonic hazards.

The objectives are fulfilled through a case study and for that the Pathar Pratima block which is one of the most cyclone risk community development blocks (Fig. 2.11) in Indian Sundarbans was selected for the study and in the case of socio-economic condition, the block is one most backward CD block and demographically the most vulnerable people are here (Pathar Pratima (Mandal et al. 2022).

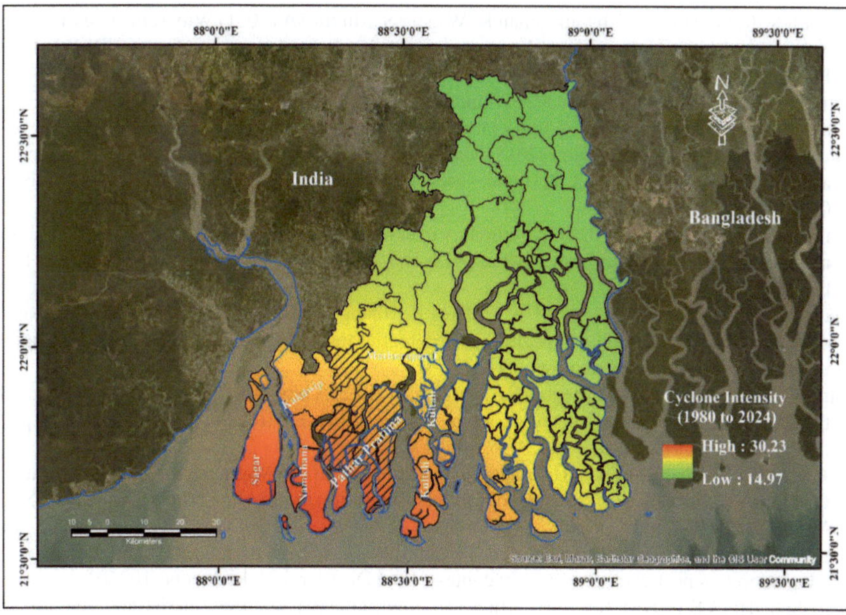

Fig. 2.11 Cyclone intensity of the Sundarbans (India)

References

Adams H (2016) Why populations persist: mobility, place attachment and climate change. Popul Environ 37(4):429–448

Aria M, Cuccurullo C (2017) Bibliometrix: an R-tool for comprehensive science mapping analysis. J Inform 11(4):959–975

Barassi M, Ercolani M, Herrerias M, Jin Z (2018) Climate anomalies and migration between Chinese provinces: 1987–2015. Energy J 39:123–144. https://doi.org/10.5547/01956574.39.SI1.merc

Bamel UK, Pandey R, Gupta A (2020) Safety climate: systematic literature network analysis of 38 years (1980-2018) of research. Accid Anal Prev 135:105387

Becerra Ruiz MJ, Pimentel MA, De Souza EB, Tovar GI (2020) Geospatiality of climate change perceptions on coastal regions: a systematic bibliometric analysis. Geogr Sustain 1(3):209–219. https://doi.org/10.1016/j.geosus.2020.09.002

Biswas B, Sultana Z, Priovashini C, Ahsan MN, Mallick B (2021) The emergence of residential satisfaction studies in social research: a bibliometric analysis. Habitat International 109:102336

Ellegaard O (2018) The application of bibliometric analysis: disciplinary and user aspects. Scientometrics 116(1):181–202. https://doi.org/10.1007/s11192-018-2765-z

Fu HZ, Waltman L (2022) A large-scale bibliometric analysis of global climate change research between 2001 and 2018. Clim Chang 170(3):1–21. https://doi.org/10.21203/rs.3.rs759304/v1

Ghosh T (2021) Climate change and migration in Sundarbans: a threat to human security. J Clim Stud

Guha I, Roy C (2016) Climate change, migration and food security: evidence from Indian Sundarbans. Int J Theor Appl Sci 8(2):45–49

Guo YM, Huang ZL, Guo J, Li H, Guo XR, Nkeli MJ (2019) Bibliometric analysis on smart cities research. Sustainability (Switzerland) 11(13):3606. https://doi.org/10.3390/su11133606

Halder NK, Merchant A, Misbahuzzaman K, Wagner S, Mukul SA (2021) Why some trees are more vulnerable during catastrophic cyclone events in the Sundarbans mangrove forest of Bangladesh? For Ecol Manag 490:119117
Hoffmann R, Šedová B, Vinke K (2021) Improving the evidence base: a methodological review of the quantitative climate migration literature. Glob Environ Chang 71:102367
Hou Q, Mao G, Zhao L, Du H, Zuo J (2015) Mapping the scientific research on life cycle assessment: a bibliometric analysis. Int J Life Cycle Assess 20(4):541–555. https://doi.org/10.1007/s11367-015-0846-2
Hualin X, Zhang Y, Duan K (2020) Evolutionary overview of urban expansion based on bibliometric analysis in web of science from 1990 to 2019. Habitat Int 95:102100. https://doi.org/10.1016/j.habitatint.2019.102100
Huang X, Yang Z, Zhang J, Wang R, Fan J, Zhang H, Huang H (2022) A bibliometric analysis based on web of science: current perspectives and potential trends of SMAD7 in oncology. Frontiers in cell and developmental biology 9:712732
Islam MM, Chowdhury MAM, Begum RA, Amir AA (2022) A bibliometric analysis on the research trends of climate change effects on economic vulnerability. Environ Sci Pollut Res 1:16. https://doi.org/10.1007/s11356-022-20524-3
Jha CK, Gupta V, Chattopadhyay U, Amarayil Sreeraman B (2018) Migration as adaptation strategy to cope with climate change: a study of farmers' migration in rural India. Int J Clim Chang Strat Manag 10(1):121–141
Khalil MB, Jacobs BC, McKenna K (2021) Linking social capital and gender relationships in adaptation to a post-cyclone recovery context. Int J Disaster Risk Reduct 66:102601
Khan MA, Ho YS (2012) Top-cited articles in environmental sciences: merits and demerits of citation analysis. Sci Total Environ 431:122–127
Khavarian-Garmsir AR (2023) A systematic review of shrinking cities literature: lessons from the past and directions for the future. Int Plan Stud 28(3-4):219–238
Kim BJ, Jeong S, Chung JB (2021) Research trends in vulnerability studies from 2000 to 2019: findings from a bibliometric analysis. Int J Disaster Risk Reduct 56:102141. https://doi.org/10.1016/j.ijdrr.2021.102141
Liang B, Shi G, Sun Z, Babul H, Zhou M (2023) Evolution trend and hot topic measurement of climate migration research under the influence of climate change. Front Ecol Evol 11:1118037
Liu W, Wang J, Li C, Chen B, Sun Y (2019) Using bibliometric analysis to understand the recent progress in agroecosystem services research. Ecol Econ 156:293–305. https://doi.org/10.1016/j.ecolecon.2018.09.001
Lu X, Wrathall DJ, Sundsøy PR, Nadiruzzaman M, Wetter E, Iqbal A et al (2016) Detecting climate adaptation with mobile network data in Bangladesh: anomalies in communication, mobility and consumption patterns during cyclone Mahasen. Clim Chang 138:505–519. https://doi.org/10.1007/s10584-016-1753-7
Mallick B, Rogers KG, Sultana Z (2021) In harm's way: non-migration decisions of people at risk of slow-onset coastal hazards in Bangladesh. Ambio 51(1):114–134. https://doi.org/10.1007/S13280-021-01552-8
Mandal P, Maiti A, Paul S, Bhattacharya S, Paul S (2022) Mapping the multi-hazards risk index for coastal block of Sundarban, India using AHP and machine learning algorithms. Trop Cyclone Res Rev 11(4):225–243
Mandal P, Roy P, Maiti A, Yadav A, Bhattacharya S, Paul S (2024) Assessing the interplay of environmental change, cyclonic hazards, and their ramifications on agriculture in Indian Sundarban. Ocean Coast Manag 251:107080
Maretti M, Tontodimamma A, Biermann P (2019) Environmental and climate migrations: an overview of scientific literature using a bibliometric analysis. Int Rev Sociol 29(2):142–158. https://doi.org/10.1080/03906701.2019.1641270
Mazumdar S, Mazumdar PG, Kanjilal B, Singh PK (2014) Multiple shocks, coping and welfare consequences: natural disasters and health shocks in the Indian Sundarbans. PLoS ONE 9(8):e105427

Mijani N, Shahpari Sani D, Dastaran M, Karimi Firozjaei H, Argany M, Mahmoudian H (2022) Spatial modeling of migration using GIS-based multi-criteria decision analysis: a case study of Iran. Trans GIS 26(2):645–668

Mistri A (2019) Is the migration from Indian Sundarban an environmental migration? Investigating through sustainable livelihood approach (SLA). Asian Profile 47(3):195–219

Milán-García J, Caparrós-Martínez JL, Rueda-López N, de Pablo Valenciano J (2021) Climate change induced migration: a bibliometric review. Glob Health 17(1):74

Mondal M, Biswas A, Haldar S, Mandal S, Mandal P, Bhattacharya S, Paul S (2022a) Rural livelihood risk to hydro-meteorological extreme events: empirical evidence from Indian Sundarban applying IPCC-AR5 and DEMATEL methodology. Int J Disaster Risk Reduct 77:103100

Mondal M, Biswas A, Bhattacharya S, Paul S (2022b) Community readiness to cyclone induced multi-hazards: evidence from villages of Indian Sundarban. Saf Extrem Environ 4(3):193–210

Mondal M, Biswas A, Mandal S, Bhattacharya S, Paul S (2023) Developing micro level resilience index for Indian Sundarban adopting resilience indicators for measurement and analysis (RIMA) methodology. Geosyst Geoenviron 2(1):100129

Moscoso P, Peck M, Eldridge A (2018) Systematic literature review on the association between soundscape and ecological/human wellbeing. Peer J Preprints 6:e6570v2. https://doi.org/10.7287/peerj.preprints.6570v2

Pachauri RK, Reisinger A (2007) IPCC fourth assessment report, vol 2007. IPCC, Geneva, p 044023

Patnaik U, Narayanan K (2015) How effective are coping mechanisms in securing livelihoods against climatic aberrations? Evidences from rural India. Int J Clim Chang Strat Manag 7(3):359–374

Rana IA (2020) Disaster and climate change resilience: a bibliometric analysis. Int J Disaster Risk Reduct 50:101839. https://doi.org/10.1016/j.ijdrr.2020.10183947

Sarkhel P, Mukherjee A (2021) Land acquisition, markets and political networks: evidence from the Indian Sundarbans. J South Asian Dev 16(2):194–219. https://doi.org/10.1177/097317412110248

Sengupta A, Samanta G (2022) Understanding immobility of a highly vulnerable coastal village in the Indian Sundarban. Reg Environ Chang 22(3):90

Shukla R, Agarwal A, Gornott C, Sachdeva K, Joshi PK (2019) Farmer typology to understand differentiated climate change adaptation in Himalaya. Sci Rep 9(1):1–12. https://doi.org/10.1038/s41598-019-56931-9

Silchenko D, Murray U (2023) Migration and climate change—the role of social protection. Clim Risk Manag 39:100472

Wang Z, Zhao Y, Wang B (2018) A bibliometric analysis of climate change adaptation based on massive research literature data. J Clean Prod 199:1072–1082. https://doi.org/10.1016/j.jclepro.2018.06.183

Chapter 3
Cyclone-Induced Coastal Vulnerability and Human Migration Risk in Sundarbans

Abstract Climate-induced migration has emerged as a critical consequence of environmental changes, particularly in vulnerable coastal regions like the Indian Sundarbans. This study investigates the multifaceted dimensions of cyclone-induced migration risk by integrating socio-economic, infrastructural, agricultural, and climatic vulnerability factors. The research is centered on the Pathar Pratima CD Block of South 24 Parganas, West Bengal—a region frequently battered by severe cyclones such as Aila (2009), Amphan (2020), and Yaas (2021). Utilizing a composite index methodology based on the IPCC AR4 framework, the study incorporates 28 indicators categorized under exposure, sensitivity, adaptive capacity, infrastructure availability, and agricultural vulnerability. Through spatial analysis in ArcGIS, migration risk zones were mapped, revealing that approximately 60% of villages fall under moderate to very high migration risk, especially those near the coastline. The findings demonstrate that cyclone frequency, saline water inundation, low adaptive capacity, and poor infrastructure collectively escalate migration risks. Case studies of three villages—Gobardhanpur (coastal), Kuemuri (riverine), and Purna Chandipur (inland)—corroborate the model's validity, showing higher migration rates in coastal areas due to recurrent cyclonic shocks and agrarian distress. Significant positive correlations were found between migration risk and variables like livelihood crises and exposure, while adaptive capacity and infrastructure availability were negatively correlated. The study further underscores the notion of "trapped populations" who are unable to migrate despite high vulnerability, due to socio-economic constraints. This research offers a robust methodological framework to assess and spatially map cyclone-induced migration risk. It provides actionable insights for policymakers, planners, and NGOs aiming to develop resilience strategies for climate-vulnerable populations. Ultimately, it emphasizes the urgent need to address the intertwined vulnerabilities to mitigate forced displacement and enhance adaptive capacity in the face of escalating climate threats.

Keywords Migration · Hazards · Potential risk · Adaptive capacity · IPCC · Livelihood

P. Mandal and S. Paul, *Climatic Vulnerability and Migration: Evidence from Indian Sundarbans*, Springer Geography, https://doi.org/10.1007/978-3-032-05858-4_3

3.1 Introduction

Global migration has always been a significant aspect of human history, and some of these movements are influenced by climatic pressures and environmental change. Environmental migration is defined by the International Organization for Migration (IOM) as human mobility caused by abrupt or gradual environmental changes that adversely affect people's living conditions and force them to leave their usual homes either voluntarily or involuntarily within their own countries or abroad either temporarily or permanently (IOM 2011). Climate migration, which is a subset of environmental migration, is the displacement of people as a result of climate-related events, dangers, and disasters. Environmental changes interact with climate change to endanger food, economic, and livelihood security, leading to migration (Schoonover et al. 2021; IPCC 2014a).

The worst of all natural disasters globally, tropical cyclones are responsible for around 64% of all fatalities (Lin et al. 2012). Around 20,000 people are killed each year, and 80–100 tropical cyclones around the world cause an annual economic loss of $6–7 billion (Shultz et al. 2005). In the Bengal delta, cyclones are a highly common form of nature. The North Indian Ocean (NIO) basin is interesting because it accounts for 7% of all global cyclonic episodes, which is approximately five times more than the Arabian Sea (Mondal et al. 2022). Located in the most northern part of the Gulf of Bengal, the Indian Sundarbans is one of the world's most vulnerable deltas. A drainage system and mangroves surround it as well (Ghosh and Mistri 2021, 2022). Because of the sea surface temperature (SST), which generates upright vortices, the Bay of Bengal basin is well-suited to producing tropical cyclones (TCs) in significant numbers (Sahoo and Bhaskaran 2016). Recent years have seen numerous intense cyclones impact coastal Bengal with heavy rainfall, including *Fani* (May 2019), *Bulbul* (November 2019), *Amphan* (May 2020), and *Yaas* (May 2021) (Wahiduzzaman et al. 2017; Uddin et al. 2019; Bhardwaj et al. 2019; Mondal et al. 2021). The Sundarbans' coastal residents are at considerable risk from the Bay of Bengal's rising sea surface temperature (SST), particularly during the pre-monsoon season when the temperature frequently reaches 30–32 °C, which is conducive to the development of depressions and cyclonic storms (Kumar et al. 2016; Ghosh and Mistri 2022). With the exception of deep depressions and depressions, the Bay of Bengal basin experienced 121 tropical cyclones (Mondal et al. 2022) over the past 50 years (1971–2020). Coastal farming activities are negatively impacted when saline water incursion inundates agricultural land and raises the soil salinity (Ghosh and Mistri 2020, 2021). Hazards along the coast have a very dynamic nature and are intimately connected to one another. Risk is created when vulnerabilities and hazards combine. Regional environment and people may be exposed to risks, but how well the environment and people are able to defend themselves from such hazards determines how susceptible the region and its inhabitants are (Pourghasemi et al. 2020a, b; UNDP 2004). The Intergovernmental Panel on Climate Change (IPCC) defines this strategy as a cumulative form of vulnerability, combining exposure, sensitivity, and adaptive capability (IPCC 2014a, b).

The terms "environment refugees," "climate refugees," and "climate migrants" have dominated research of extreme climate change-related events and forced migration for more than three decades, although some of these terms lack critical examination (Su 2014). Indeed, Baldwin et al. (2014) show that environmental or climatic "refugees" remain a hypothetical possibility, but neither exists nor is precisely characterized. Despite disagreement among experts regarding the relationship between climate change, the environment, and migration, the International Organization of Migration (IOM) have already issued warnings that human migration and displacement may be the most serious consequence of climate change and may intensify environmental migrations. The number of persons affected by natural disasters and risks since 2010 has increased to approximately 1.5 billion (Alizadeh et al. 2018; Karimi et al. 2021; Mondal et al. 2021; Ghosh and Mistri 2022). Climate change also threatens the agricultural and fishing industries because of rising sea levels, coastal erosion, and land submersion (Brown and Nicholls 2015; Mandal et al. 2018). Those who live along the coast suffer severe losses to their lives and means of subsistence due to the cumulative effects of these several risks, forcing them to leave these areas (New et al. 2006; Le et al. 2007). Nonetheless, a specific hazard's physical and socio-economic sensitivity might vary and have an impact on society and the environment on a larger scale (Mondal et al. 2021). According to the New Displacement World report from 2019, a total of 33.4 million people have been uprooted, of which 24.9 million were uprooted due to disasters. Weather-related disasters forced the displacement of over 23.9 million people, of which 13 million people were displaced by storms (11.9 million by cyclones, hurricanes, typhoons, and 1.1 million by other storms). In India, the Internal Displacement Monitoring Center (IDMC) shows that such a vulnerable population has been uprooted, with the majority of the uprooting caused by floods (52.76%) and storms (45.98%), with the remainder being caused by drought (1.26%). Due to the country's high levels of social and economic vulnerability, maximum population density, and growing hazard intensity, five million additional disaster-related displacements have occurred in India. Ultimately, according to estimates by IDMC, disasters in India are anticipated to cause roughly 2.3 million people to be displaced each year (Council–NRC, N. R 2014).

Continuous natural hazards have a negative influence on people's resources, income, and ability to resist migration in coastal areas, hence reducing their capacity for coping with stress (Siddiqui 2014). The coastal Sundarbans of West Bengal (South and North 24 Parganas), India, has been particularly vulnerable to floods and cyclones, which have led to a variety of human mobility issues, including emigration and relocation. In 2009, cyclone *Aila* had a negative impact on this area, and very recently, *Amphan* in May 2020 had a negative impact (Fig. 3.1). Around 60 million people were impacted, and at least 80 people died in West Bengal alone, according to the OCHA (26 May 2020) report. An attempt could be made in this situation to research the vulnerability and susceptibility to migration risk associated with tropical storms. These days, it is exceedingly difficult for researchers to gauge the intensity of migration in spatial scenarios mostly because of climate risks. Because there is no appropriate database, no clear methodology, and the environment has a significant impact on economic, political, social, and demographic factors, it is more difficult to determine climate-induced migration (Black et al. 2011). So, identifying the possibility of climate-induced migration by combining the coastal vulnerability index with IPCC AR4-2007 (Adger et al. 2007) methodology, including agriculture vulnerability, and infrastructure availability can help to lessen the problem's intensity (Ghosh and Mistri 2022). There hasn't been much research done on crop damage, additional production losses, how agricultural susceptibility due to cyclonic risks increases the probability of population migration, or how these factors interact with the economic situation of the afflicted household.

With regard to previous agricultural risks (caused by the cyclones of *Bulbul*, 2019, *Amphan*, 2020, and *Yaas* in 2021), as well as current coastal risks (Rakib et al. 2019), this study provides a novel methodological approach. In order to assess the migration risk of the people in coastal areas, a method for estimating the same has been presented. It takes into account infrastructural accessibility, agricultural vulnerability to cyclones, and sensitivity to cyclones, as well as adaptive capability, which may be a key coping mechanism in any unfavorable situation. The specific objectives of the study are as follows: (a) to assess the nature of exposure, sensitivity, and socio-economic adaptive capacity at village level due cyclonic activity, (b) to assess the agricultural vulnerability and availability of infrastructure, (c) to quantify the vulnerability integrated adaptive capacity and spatial pattern of migration risk.

Fig. 3.1 Coastal Sundarban Schools remain temporarily closed following 35 days of Cyclone Amphan 2020, putting children's safety at risk. *Source* By authors on 27-06-2020

Three research questions were formulated to assess the above-mentioned objectives, which are as follows:

(1) what is the present scenario of socio-economic conditions in the coastal community?
(2) How does crop land inundation production loss increase the livelihood crisis?
(3) How does integrated adaptive capacity influence coastal habitats toward migration risk?

3.2 Research Methods

3.2.1 Study Area

In South 24 Parganas district of West Bengal, India's Pathar Pratima CD Block is situated in the southernmost part of the country and is bordered by Bangladesh in the east, the Hooghly River in the west, and the Gulf of Bengal in the south. The study area is 484.47 km^2 and is located between 88°38′E and 88°30′E and 21°38′N and 22°02′N. Geographically, the Pathar Pratima CD Block of the Kakdwip subdivision, which is surrounded by the Namkhana, Kulpi, Mathurapur, and Kakdwip CD blocks, covers an area of 484.47 km^2 (Fig. 3.2). The entire research region, which spans fifteen (15) Gram Panchayats, one (01) Panchayat Samity, and eighty-six (86) populated villages, is located at the lower deltaic plain at a height of 4 m above mean sea level. Due to the block's deltaic location, the Ganga River's distributaries have considerably cut across it (Pal and Paul 2021). The river network serves as this region's main transportation system and a way of survival for its residents (fishing). The deltaic region's drainage system is tidal in nature and has year-round access to a significant amount of water. As a result, erosion of river banks and river embankment breaches are common in this area. Cyclones Sidr and Aila affected around 2.3 million people living in the Indian Sundarbans (Giri et al. 2022). Sea levels have risen, rainfall patterns have changed, and the frequency and severity of extreme weather events have increased for the residents of the Indian Sundarban (Dubey et al. 2017). Results of cyclonic storm surge flooding, agricultural areas are submerged by saline water (Chand et al. 2012; Rahman et al. 2020). Due to numerous strengthened cyclones making landfall in the Sundarban region or nearby areas over the past 5 years (2018–2022), the situation has significantly deteriorated. Cyclones like the Fani, Bulbul, Amphan, and Yaas, which only have a 6–12 months window before making landfall, are a serious setback for this area.

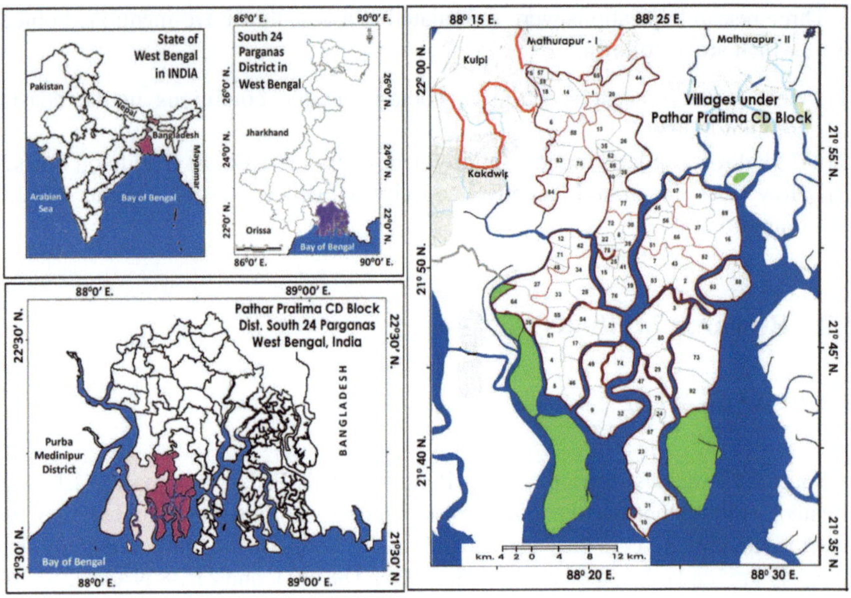

Fig. 3.2 Location of the study area

3.2.2 Selection of Indicators and Data Collection

In the geographical setting of frequently multiple threats, migration decisions take a variety of socio-economic considerations into consideration (Martin et al. 2014). Rural migration occurs as a result of the structure and dangers of rural livelihoods, depending on agriculture (Mallick and Vogt 2014). Coastal population displacement is significantly impacted by crop failures and the resulting financial losses from flooding (Gray and Mueller 2012). Mallick and Vogt (2014) found that the majority of those who fled following *Aila* (2009) all resided in mud dwellings (both wall and roof), with 50% of them living below the poverty line. Most commonly, economic hardship is the major reason behind the decision to migrate to the village. According to studies, marginal workers in India can only find employment in agricultural land for a maximum of six months per year (Mallick and Vogt 2014). Martin et al. (2014) found in his study that migration is viewed as a replacement for increasing household income. Local settlements were abruptly relocated in 2009 as a result of Cyclone Aila and the storm surge it created (Martin et al. 2014). Agricultural growth was hindered throughout the year due to a lack of irrigation facilities, which depicts stories of helplessness in the face of uncertainty. Yet, the respondents noted the strength of human agency in taking appropriate adaptive action, such as migration (Martin et al. 2014). Regarding adaptive capacity or what can be done given the resources' accessibility and availability (Grothmann and Patt 2005). In their study, Mondal et al. (2021) demonstrated that socio-economic illiteracy increases the probability of cyclone disasters for coastal residents. A higher level of vulnerability is mostly caused by the proportion of females and children in a home. According to Perch-Nielsen et al. (2008), a number of adaptation options exist in the impacted areas that offer an alternative to migration, which led them to draw the conclusion that, at the very least, floods "will not likely be a primary mechanism by which climate change would cause widespread migration."

As the climate migration index was created in a spatial dimension, the paper's primary goal has already been covered. To carry out this strategy, five categories of the component—cyclonic risks, coping capacity, adaptive capacity, livelihood crises, as well as infrastructure accessibility—were established using a variety of empirical variables that were created by numerous academics after extensive empirical research on climate migration. A secondary database was utilized during the entire investigation. The information came from a variety of official sources, including the Socio-economic Caste Census (SECC, 2011), house listing primary census abstract (HLPCA, 2011), Primary census abstract (PCA, 2011), District census handbook (DCHB, 2011), and all the details have been framed in Table 3.1.

3.3 Estimation of Human Migration Risk

3.3.1 Hypothesized Theory

Over 170 million of India's 1.4 billion inhabitants who live in coastal areas are under risk from salinization, strong tropical cyclones, embankment breaching, and climate change. The most recent instance of this vulnerability was in May 2020, when Cyclone *Amphan*, the worst cyclone to hit the Bay of Bengal in several decades, forced several million people to flee. India suffered 235 km^2 of land loss due to coastal erosion between 1990 and 2016, putting people's homes and livelihoods at peril. As a last resort, the government had to step in to help people flee to safer locations. In South Asia, the monsoon rains are the greatest in absolute terms, and the Internal Displacement Monitoring Centre estimates that between 2008 and 2018, over 3.6 million Indians were displaced annually, largely due to flooding from those storms. Theoretically, the multi-hazards principally brought on by tropical cyclones can be overcome by migration risk (Fig. 3.3).

Emerging scientific data suggests that man induced climate change is influencing the pattern of migration and displacement more dramatically along the Sundarbans coastal area, even though it is challenging to define and quantify the direct causation of climate change on human mobility. Socio-economic vulnerabilities, decreased agricultural output, lower crop yields, and damage to livelihood capitals all have the cumulative effect of raising the risk of mass migration from the susceptible coastal region to the interior of the state or to neighboring states. Natural disasters and their associated hazards, whose effects might result in socio-economic vulnerabilities, increase in poverty, decreased possession of assets, and loss of livelihood options

Table 3.1 Variables used to assess vulnerability and migration risk

Indicators	Unit of measurement		Literature reviewed	Data source
Sensitivity				
Population density	(S_1)	Person km^{-2}	Ghosh and Mistri (2022), Mallick (2013), Mondal et al. (2022), Martin et al. (2014)	DCHB (2011), SECC (2011), HLPCA (2011), SECC (2011), PCA (2011)
% of HHs having only one room with kucha wall and roof	(S_2)	%		
Agricultural dependency	(S_3)	%		
Average household size	(S_4)	Number		
Sex ratio	(S_5)	Number		
Youth dependent population (0–15 years)	(S_6)	%		
Socially disadvantaged people	(S_7)	%		
Amount of land holding	(S_8)	Number		
Deprived households	(S_9)	%		
Marginal workers	(S_{10})	%		
Adaptive capacity				
Literacy rate	(AD_1)	%	Rakib et al. (2019), Martin et al. (2014), Mondal et al. (2022), Mallick (2013)	PCA (2011), SECC (2011), HLPCA (2011), DCHB (2011)
HHs with Rs. 10,000 or more income	(AD_2)	%		
HHs without non-permanent structures	(AD_3)	%		
Area under irrigation	(AD_4)	%		
Crop diversity ratio	(AD_5)	Number		
HHs without electricity	(AD_6)	%		
HHs without a safe drinking water facility	(AD_7)	%		
HHs without sanitation	(AD_8)	%		
Livelihood crisis				
Farmland damaged by the cyclonic activity	(LC_1)	%	Rakib et al. (2019)	ADO, Pathar Pratima CD Block
Production loss (in MT)	(LC_2)	Number		

(continued)

Table 3.1 (continued)

Indicators	Unit of measurement		Literature reviewed	Data source
Monetary loss (in Lac Rs.)	(LC_3)	Number		
Per-capita farmer monetary loss	(LC_4)	Number		
Infrastructure availability				
Distance of primary health centre (PHC)	(IA_1)	Km	Mondal et al. (2021), Hoque et al. (2021), Martin et al. (2014), Mallick (2013)	DCHB (2011)
Distance of district hospital (DH)	(IA_2)	km		
Distance of market	(IA_3)	km		
Distance of all-weather road	(IA_4)	km		
Distance of flood/cyclone shelter	(IA_5)	km		

Note HH means Households

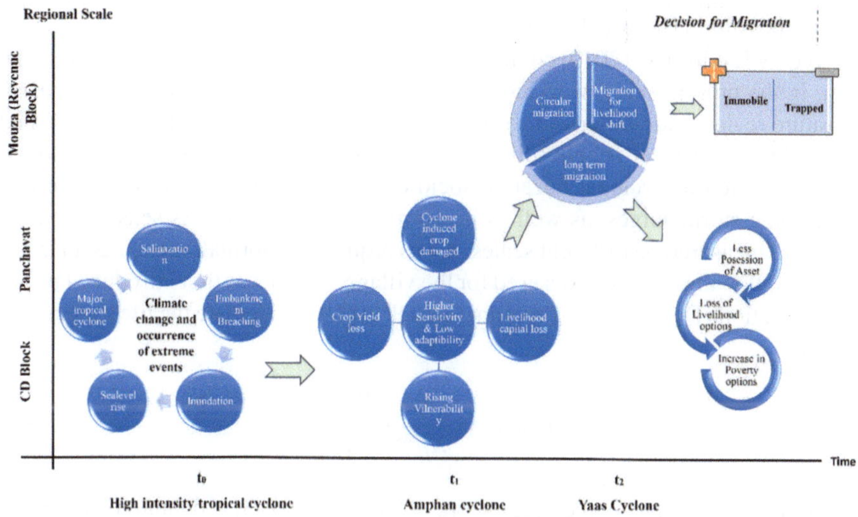

Fig. 3.3 Conceptual frameworks for coastal migration risk. *Source* By authors after modification of Black et al. (2011)

with realized adaptive capacity, are a primary cause of worry for coastal population migration risk. Several families are migrating or have already relocated to other places as a result of population displacement caused by multiple threats brought on by the climate in coastal West Bengal. It is very likely that the current pattern of relocation along India's coastal regions, including the Sundarbans, would become more pronounced due to climate change and catastrophic occurrences.

3.3.2 Formulation of Index

In this study, the risk of human migration from the Pathar Pratima block is evaluated by taking into account exposure, sensitivity, adaptability, agricultural vulnerability, and available infrastructural facilities. Potential effects that are positively connected to vulnerability are made up of exposure and sensitivity together. Yet, studies show that through improving coping skills, adaptive capacity boosts resilience (Ghosh and Mistri 2021; Mondal et al. 2021). For the purpose of measuring the migration risk index in the current study, a total of 28 variables were chosen. According to IPCC AR5, the risk is the potential for repercussions to identify the diversity of values and that something valuable is at stake with an unknown outcome. It is frequently portrayed as the result of the likelihood that hazardous occurrences or trends will occur along with their effects. Risks related to the effects of climate change are the main uses of the word risk (Das et al. 2020).

To quantify migration risk at the household level in the current study, the aforementioned indices were included in a composite index approach. The index methods must be trustworthy, tenable, and substantial when comparing the estimated scores of various components with the scale in order for the study's results to be as acceptable as possible. The equal-weighted index suggested by Sullivan (2002) and also utilized by Rakib et al. (2019) could not be employed because there were no baseline data. In the present study, working indicators were developed through an intensive literature review. Additional indicators were selected based on field observation and discussions with local people (during the pilot survey, 2022). Most of the indicators are related to cyclonic hazards, socio-economic resilience, crop damage due to cyclonic storm surges, as well as infrastructure availability. As selected indicators were measured on different scales, it was required to normalize each as an index value and flowing equation was used for the village, which was first introduced by the United Nations Development Program (UNDP 2007) to calculate the life expectancy index:

$$\text{index}_{ij} = \frac{X_{ij} - X_j^{\min}}{X_j^{\max} - X_j^{\min}} \tag{3.1}$$

where index ij represents the index score of i-th (i = 1, 2, 3, 4, 5) union and j-th (j = 1, 2, 3...49) indicator. X_{ij} is the original value of i-th union and j-th indicator, and $X_j^{\min}$, and $X_j^{\max}$ are the minimum (lowest) and the maximum (highest) values of the j-th indicator, respectively. All the possible normalized scores will range from 0 to 1. After the normalization of each indicator, using Eq. (3.2), the average is to obtain the scores of each component (Hazard, Sensitivity, adaptive capacity, agriculture vulnerability, and infrastructure availability.

$$C_i = \sum_{j=1}^{m} \text{index}_{ij} / m \tag{3.2}$$

where C_i indicates the component score of union i and m represents the total number of indicators of each component. Each component score is spatially analyzed by mapping environment and done by Arc GIS software, and the overall result of each village was classified into five classes as very high (5), high (4), moderate (3), low (2), and very low (1).

After the estimation of the component C_{ij} of each village, the study estimates socio-economic vulnerability with the help of IPPC AR4 methodology:

$$\mathrm{SEV} = f(\mathrm{E, S, AC}) \tag{3.3}$$

where SEV = Socio-economic vulnerability, E = Exposure, S = Sensitivity, and AC = Adaptive Capacity. It has to be noted that the adaptive capacity of a system determines the vulnerability by modulating exposure and sensitivity (Adger et al. 2007).

Before the estimation of vulnerability, the study first estimated potential impact as the combination of exposure and sensitivity, where the adaptive capacity is not included (Hoque et al. 2021). And the equation like as

$$\mathrm{PI} = \mathrm{E} \times \mathrm{S} \tag{3.4}$$

where PI is equal to potential impact and E and S are already described in Eq. (3.3).

3.3.3 Estimating Migration Risk

Finally, migration risk is estimated by the amalgamation of Socio-economic Vulnerability, agricultural vulnerability, and infrastructure vulnerability. Numerous studies suggested that linear or equal weight is not a suitable method for such a kind of estimation (Rakib et al. 2019; Ghosh and Mistri 2021; Mondal et al. 2021). The study, as per its objective, considers that cyclonic exposure is the most dominant vulnerability factor, followed by adaptive capacity. Infrastructure availability is helpful for short-term relief, but cannot provide them a long-time livelihood opportunity. Cyclonic vulnerability is directly linked with socio-economic vulnerability and receives maximum weightage for the construction of migration risk indicators. Among the three, the study allotted 3/3 = 1 weightage for socio-economic vulnerability which is directly linked with cyclonic vulnerabilities, 2/3 = 0.66 weightage for adaptive capacity and 1/3 = 0.33 weightage for infrastructure vulnerability. Thus, the study constructed migration risk as follows Eq. (3.5).

$$\mathrm{MR{:}\ SEV} + (\mathrm{AV} \times 0.66) + (\mathrm{IV} \times 0.33) \tag{3.5}$$

where MR means Migration risk, AV is the agriculture vulnerability, and IV is the Infrastructure vulnerability.

3.3.4 Status of Infrastructure and Agricultural Vulnerability

Infrastructure vulnerability is estimated by the study opted methodology. Villages located in closure proximities are less vulnerable in comparison to others. The quantification is done by adopting four values, like greater than 10 km, between 5 and 10 km, less than 5 km, and the binary response (yes) about the present location within the considered mouza. Reminding the sense of infrastructure availability, the study allotted weights among the four categories with minimum exposure on greater than 10 km (1/4 = 0.25), for between 5 and 10 km (2/4 = 0.50), for less than 5 km (3/4 = 0.75), and lastly for living within the mouza (4/4 = 1). Then the study estimated the infrastructure availability score following Eqs. (3.1) and (3.2). Thus, as per the study on infrastructure availability, the maximum value indicates the greater availability of infrastructure (Fig. 3.4). But we have the objective to estimate the infrastructure vulnerability, which captures the reverse groups as designed in terms of available infrastructure. Hence, considering such reverse dimensions between infrastructure availability and infrastructure vulnerability, the study uses Eq. (3.6) to capture infrastructure vulnerability.

$$\text{IV}: 1 - \text{IA} \tag{3.6}$$

where

IV is the infrastructure vulnerability, and IA is the infrastructure availability.

The study uses four variables to estimate the agriculture vulnerability, and these are crop (paddy) area damages, paddy production loss, total monetary loss, and per capita monetary loss. The village-level data source of crop-related loss and monetary

Fig. 3.4 A person from the coastal area told us that his son had to leave the area in quest of better employment possibilities after a cyclone badly wrecked his home. The money his son sent home was essential in helping the family build a concrete home that would give them a safer and more long-lasting place to live in the event of another calamity. *Source* By authors on 07-07-2023

Fig. 3.5 The agricultural land behind her home was planted in 2019, according to a local woman. Nevertheless, Cyclone Bulbul's flooding totally destroyed the standing crop, which had reached maturity. In addition to affecting family food security, this unexpected loss also interfered with their seasonal revenue. *Source* By authors on 07-07-2023

loss due to the last three intensive cyclones, such as *Bulbul* (2019), *Amphan* (2020), and *Yaas* (2021) (Fig. 3.5) is used by the study, and it is collected from the Agricultural Development Office (ADO, Pathar Pratima). The study deflated the three data sets in terms of the price level of 2022 and then took the average to estimate the agricultural vulnerability. The per capita monetary loss is estimated from the total monetary loss and population in a particular village. The loss of these variables of three cyclones was averaged in all the villages, and agriculture vulnerability was calculated. The maximum value indicated the most agriculturally vulnerable village unit and vice versa.

3.4 Spatial Distribution of Socio-economic Vulnerability

Descriptive statistics of the dataset used in this study has been depicted in Table 3.2. Exposure, according to the IPCC's fourth assessment report, refers to the amount of land that is negatively impacted by cyclonic activity (Das et al. 2020). The most significant variables include things like how many cyclonic tracks have struck specific villages over the past 30 years, and other ones like the recent disastrous storm Amphan that inundated how much paddy crop land. The sum of these factors led to the development of the hazard index. Five categories, including very high exposure, high, moderate, low, and very low exposure, were created based on the final hazard scores of each village.

Table 3.2 Descriptive statistics

Variables	Mean	Standard deviation	Skewness	Range	Min.	Max.
Cyclonic hazard	3.8	0.4	0.1	1.39	3.11	4.5
Area of agricultural land inundated with percentage	35.69	6.92	−0.02	27	23.51	50.51
Population density	736.62	368.1	3.45	2483.65	0	2483.65
Only one room with Kutcha walls and Kutcha roof	60.48	14.73	−0.42	53.47	30.67	84.14
Main agricultural dependency	22.28	14.78	0.46	52.25	0.48	52.73
Sex ratio	937.94	20.46	0.5	79	900	979
0–15	30.41	2.13	1.55	11.09	27.2	38.28
SC and ST population	23.71	15.31	1.64	73.6	2.83	76.43
Net area sown (in hectares)	395.62	204.79	0.11	779.2	29.5	808.7
Share deprived household	10.09	9.59	2.38	40.1	3.15	43.25
Marginal worker	59.98	18.33	−0.24	71.37	26.2	97.57
Literacy	72.21	4.43	0.05	25.14	59.33	84.48
Rs. 10,000 or more	0.21	0.27	2.8	1.06	0.05	1.11
Permanent structure	50.63	10.77	−0.59	46.3	23.2	69.5
Total irrigated land (in hectares)	1.54	4.53	3.89	22.22	0	22.22
Crop diversity	1.94	0.96	0.14	2	1	3
Without safe drinking water	99.9	0.39	−5.38	2.2	97.8	100
Number of households having a latrine facility within the premises	54.65	11.13	−0.7	52.1	21.6	73.7
Area of agricultural land inundated with percentage	35.69	6.92	−0.02	27	23.51	50.51
Area damaged computed to 100%	50.69	10.16	0.72	36.48	36.52	73

(continued)

Table 3.2 (continued)

Variables	Mean	Standard deviation	Skewness	Range	Min.	Max.
Production loss (MT)	488.79	105.77	0.38	408.28	291.88	700.16
Money loss (Rs. Lac)	87.61	20.43	0.62	73.11	53.97	127.08
per capita loss	68,049.43	82,633.4	3.11	385,620.9	19,814.72	405,435.6
Primary health centre (numbers)	0.32	0.09	−0.87	0.3	0.1	0.4
Primary health sub centre (numbers)	3.06	1.5	−1.4	3.9	0.1	4
Regular market	2.03	1.82	0.2	3.6	0.4	4
All weather road	0.41	0.48	5.5	2.7	0.3	3
Distance of flood/ cyclone shelter	26.49	6.62	0.69	26.19	15.47	41.66

Source Computed by the Researcher

According to Fig. 3.6a, 61% of the whole area is classified as being physically exposed to high levels of risk (Fig. 3.7), compared to 20% of the total area that is highly exposed and 15% of the total area that is affected by cyclones, which together constitute a high-exposure region. Contrarily, 14% of the locations are less exposed. An area with a high to moderate risk of cyclone exposure encompasses nearly the whole southern part of the Pathar Pratima block. The villages fall under very high exposure are Gobardhanpur, Sitarampur, Buraburir Tat, Dakshin Surendraganj, Indrapur, Krishnadaspur, Rakshaskhali, Gobindapur Abad, Banashyam Nagar, Kshetra Mohanpur, Gangapur, Rakhalpur, Paschim Sripatinagar, Uttar Surendraganj, Dakshin Shibganj, Sibnagar, Purba Sripatinagar, Chhoto Banashyam Nagar, Brojaballabpur, and lastly Kuemuri mouza. These villages fall under the category of "extremely high exposure" because they were most severely impacted by cyclones in the research area over the past 30 years compared to other villages. Other villages, likely Purna Chandrapur, Biswanathpur, Srinarayanpur, Dakshin Durgapur, Dakshin Gangadharpur, Bhajna, Dakshin Raypur, Abad Gangadharpur, Gurudaspur, Dakshin Madhusudanpur, Piprakhali, Malaya, Ramdebpur, Nilerat, have very low exposure areas because of the less cyclone tracks that went through these villages, and also obviously less cropland was inundated due to the adverse effects of the last 30 years of cyclone. The northernmost part of the research region is where all of these communities are located. Given that the cyclone originates at the sea surface and moves inland, it follows that the closest village to the coast is highly exposed (Hoque et al. 2021). Around 20 out of the 87 inhabited villages in the Pathar Pratima block have very high cyclonic exposure, while 14 of the 87 inhabited villages have very low exposure (Table 3.3).

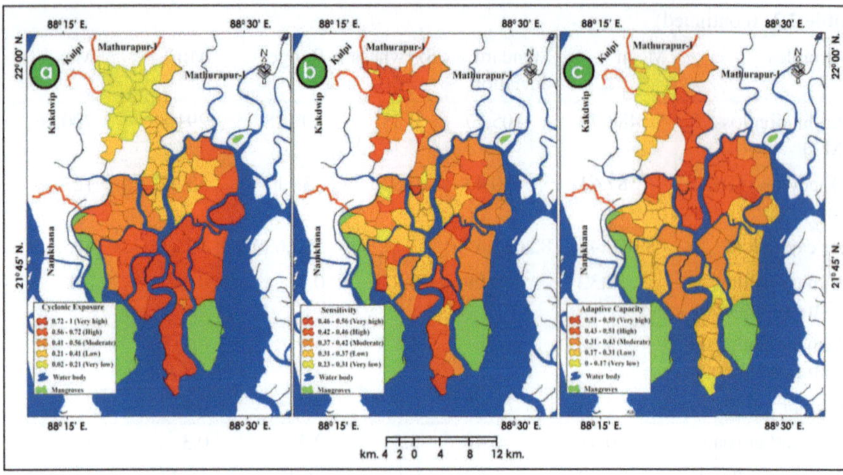

Fig. 3.6 Spatial distribution of vulnerability's variables, **a** cyclonic exposure, **b** sensitivity, and **c** adaptive capacity

Fig. 3.7 The house, composed of kutcha materials such as mud and thatch, was located precariously on the edge of the coastal bank, making it one of the most susceptible constructions in the area. Because of its location, it was immediately vulnerable to cyclonic winds and tidal surges, which increased the possibility of serious damage or total destruction during extreme weather occurrences. *Source* By authors on 19-04-2023

Table 3.3 Number and share of villages under different components

	Very high		High		Moderate		Low		Very low	
Components	Villages	% Share	Villages	% Share	Villages	% Share	Villages	% Share	Villages	% Share
Exposure	20	22.99	13	14.94	20	22.99	20	22.99	14	16.09
Sensitivity	10	11.49	20	22.99	37	42.53	16	18.39	4	4.60
Adaptive capacity	9	10.34	22	25.29	30	34.48	21	24.14	5	5.75
Agriculture vulnerability	9	10.34	18	20.69	17	19.54	18	20.69	25	28.74
Infrastructure availability	14	16.09	24	27.59	22	25.29	22	25.29	5	5.75
Potential impacts (exposure and sensitivity)	15	17.24	17	19.54	25	28.74	17	19.54	13	14.94
Socio-economic vulnerability	9	10.34	21	24.14	18	20.69	25	28.74	14	16.09
Migration risk	10	11.49	21	24.14	21	24.14	23	26.44	12	13.79

Source Computed by the Researcher

The spatial pattern of sensitivity was depicted in Fig. 3.6b and was determined by a composite index of nine independent factors, including population density, only one room with kucha walls and roof, main agriculture dependency, sex ratios of 0–15, SC and ST population, net area sown (in hectares), share deprived households, and last marginal worker. Five categories were created based on how sensitively each village scored. While individual very high and high-class shares are 11.49% and 22.99%, respectively, the proportion of villages with moderate to very high sensitivity is approximately 77.01%. The villages of Bhagbatpur, Dakshin Gangadharpur, Gobardhanpur, Banashyam Nagar, Sagar Madhabpur, Ramdebpur, Styadaspur, Abad Gangadharpur, Parbatipur, and last but not least Bhajna mouza are among the most sensitive. Figure 3.6b also shows that no such spatial pattern of sensitivity exists that would suggest a lower sensitivity scenario in the study region. Mahendra Nagar, Meherpur, Biswanathpur, and Nilerat are the least vulnerable villages, making up 4.60% of all inhabited villages, exceptionally low. In the categories of high, moderate, and low sensitivity, respectively, twenty, thirty-seven, and sixteen communities are included. Among the variables used to construct the index, sensitivity has a 47% link with the sex ratio, a 55% correlation with only one room with kucha walls and kucha roof, a 56% correlation with the 0–15 population, and a 34% correlation with the last marginal worker. As a result, these traits that indicate their backwardness are very sensitive to 11.49% of villages (Table 3.3). Additionally, the hamlet is rendered more sensitive by the presence of more kids from socially underprivileged (ST, SC) groups, by the presence of kucha walls, and by the presence of kucha fences. Women and children are a more sensitive population in coastal Sundarbans (Fig. 3.8).

The ability to manage the negative effects of extreme weather events caused by climate change is known as adaptive capacity. Here, the adaptive capacity is measured using an index form, and scores for each village's adaptive capacity are generated using data on each village's share of literacy, household income of at least Rs. 10,000 per month, the percentage of families living in permanent structures, the percentage of irrigated land, the percentage of households receiving safe drinking water, and the final number of houses with access to latrines on the property.

Figure 3.6c illustrates that the adaptive capacity of the household is low to moderate in the southern part of the research area. The majority of the northern villages have a moderate to higher potential for adaptation than the southern part. Each village's adaptable capacity score is divided into five categories, including very high, high, moderate, low, and very low adaptive ability. The percentage of villages with moderate to very high adaptable capacity is almost 70%, while the percentages of people with very high and high adaptive capacity are 10.34% and 25.29%, respectively. Gobardhanpur, Dakshin Gangadharpur, Abad Gangadharpur, Bhajna, and lastly Dakshin Madhusudanpur are the settlements that originate from the northern part of the region and have very poor adaptability. The 30 villages with a moderate ability to adapt are located in areas with very high cyclonic exposure. The village has a 5.75% share of lower adaptive capacity (Table 3.3). The low to moderate adaptation ability of the communities is caused by poor literacy, a less stable household structure, a reliance on monoculture or a lack of variety in the crops, and last

Fig. 3.8 Women and children are especially at risk in areas that are prone to cyclones, especially in households where the majority of the male members have left for employment. Women are frequently left to handle all aspects of household management, kid safety, and emergency response when adult men are absent during and after catastrophic weather disasters. This makes people more vulnerable to the negative effects of natural calamities and adds to their financial, mental, and physical load. *Source* By authors on 18-11-2023

but not least, a lack of sanitation facilities on the property. Thus, there is a positive correlation between these variables and adaptive capacity.

3.4.1 Infrastructure Availability and Agriculture Vulnerability

The final infrastructure availability score is divided into five categories. More than 55% of the area is classified as having very low to moderate infrastructure availability, and the remaining 5.75% is classified as having extremely poor infrastructure availability (Fig. 3.9a). There are a number of villages in Dakshin Gangadharpur, Meherpur, Ramdebpur, Nilerat, and Abad Gangadharpur, where infrastructure is scarce. These villages are from the northern region. These regions are described as being far from the principal health facility (Fig. 3.10), far from the typical market,

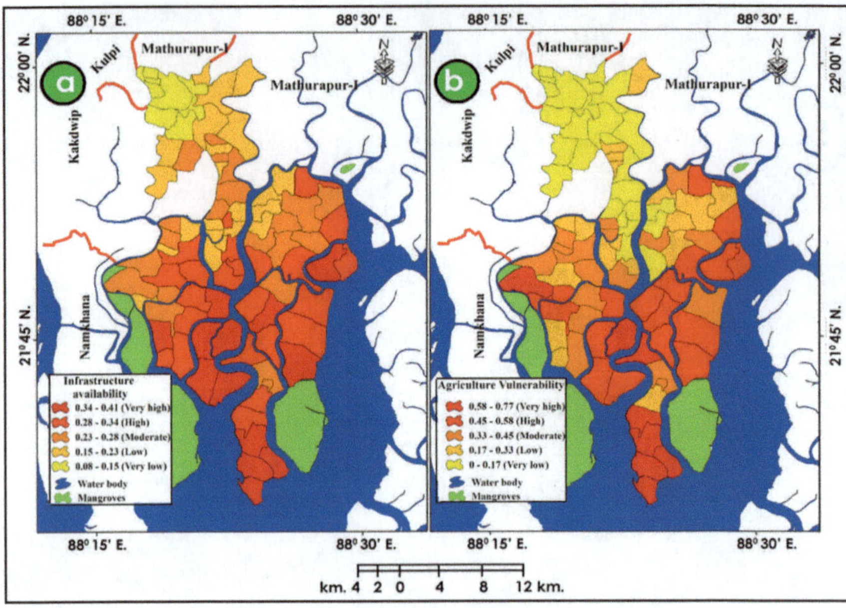

Fig. 3.9 Spatial distribution of **a** infrastructure availability and **b** agriculture vulnerability

and far from the cyclone/flood shelter. In addition, the research area's infrastructure is available in around 16.09% and 27.59%, respectively, of the study area.

The villages with the highest infrastructure availability include Gobindapur Abad, Brojaballabpur, Indrapur, Buraburir Tat, Sitarampur, Rakshaskhali, Kshetra Mohanpur, Dakshin Shibganj, Gobardhanpur, Baradapur, Sibnagar, Paschim Sripatinagar, Sridhar Nagar, and last but not least, Uttar Surendraganj. All of these communities are located in the southern region. In the region of the south, southeast, and certain villages from the southwest parts, there was a very high to high sensitivity to agriculture (Fig. 3.9b). The agriculture vulnerability score was determined by combining the crop area and production damage caused by a recent cyclone in each village separately. These variables include crop (paddy) area damage, production loss, loss of money values (rupees in laces), and the final per capita financial loss of farmers who were affected by this cyclone.

Five categories were created from the final agriculture vulnerability score. Over 50.57% of the population is in the Moderate to Very High class (Table 3.3). This indicates that close to or more than half of the area is classified as a moderate to very high agriculture risk zone. Villages that fall under this category include Gobardhanpur, Rakshaskhali, Krishnadaspur, Buraburir Tat, Sitarampur, Paschim Surendra Nagar, Dakshin Surendraganj, Indrapur, and, finally, Gobindapur Abad. Large and devastated agricultural fields, significant output losses, and significant financial losses for farmers are the characteristics of these places. Due to low crop damage, low crop yield, and financial losses to farmers, the majority of the villages' northern sections

Fig. 3.10 Primary health center at G-Plot, Indrapur, Pathar Pratima Block, South 24 Parganas. *Source* By authors on 19-11-2023

fall into the category of low to extremely low agriculture vulnerability. Furthermore, the low to very low vulnerability category encompasses almost 49% of the territory. The final five villages with minimal agricultural vulnerability are Taranagar, Dakshin Raypur, Gurudaspur, Dakshin Durgapur, and Nilerat (Fig. 3.11).

3.4.2 Potential Impact Without Integrated Adaptive Capacity

The spatial analysis derived from Fig. 3.12a reveals significant insights into the vulnerability distribution across the study region in relation to cyclonic hazards. A substantial portion of the area, amounting to 65.52%, is categorized as having moderate to very high susceptibility to cyclones. Within this segment, 17.24% (9 regions) and 19.54% (21 regions) are marked as having very high and high susceptibility levels, respectively. These vulnerable zones predominantly encompass areas such as Gobardhanpur, Buraburir Tat, Dakshin Surendraganj, Banashyam Nagar, Indrapur, Krishnadaspur, Sitarampur, Gobindapur Abad, Gangapur, Rakhalpur,

Fig. 3.11 Kharif paddy crop was totally damaged by cyclone bulbul in 2019, and also Boro paddy cultivation was destroyed by cyclone Amphan in 2020. *Source* By authors on 10-01-2024

Sibnagar, Kshetra Mohanpur, Paschim Sripatinagar, Chhoto Banashyam Nagar, and Satyadaspur.

These areas are particularly susceptible due to their demographic and socio-economic characteristics. The majority of households in these regions reside in vulnerable kucha structures (made from mud and other temporary materials), which heightens their exposure to cyclonic damage. Additionally, many residents are engaged in marginal occupations, rendering them economically sensitive to environmental disturbances. Agriculture, the primary livelihood in these areas, is highly vulnerable to cyclonic impacts, as flooding during such events often results in crop devastation, further exacerbating economic challenges.

Conversely, regions located in the northern and northwestern parts of the study area are classified as low to extremely low vulnerability zones, constituting 19.54% and 14.94% of the area, respectively (Table 3.3). These zones are less susceptible to cyclones due to their inland location, which shields them from the direct path of cyclonic winds and storm surges. Their geographical distance from the coast reduces the frequency of cyclone impacts, thereby minimizing their overall sensitivity to such natural disasters. This delineation between high and low vulnerability zones highlights the differential risk posed by cyclones within the study region.

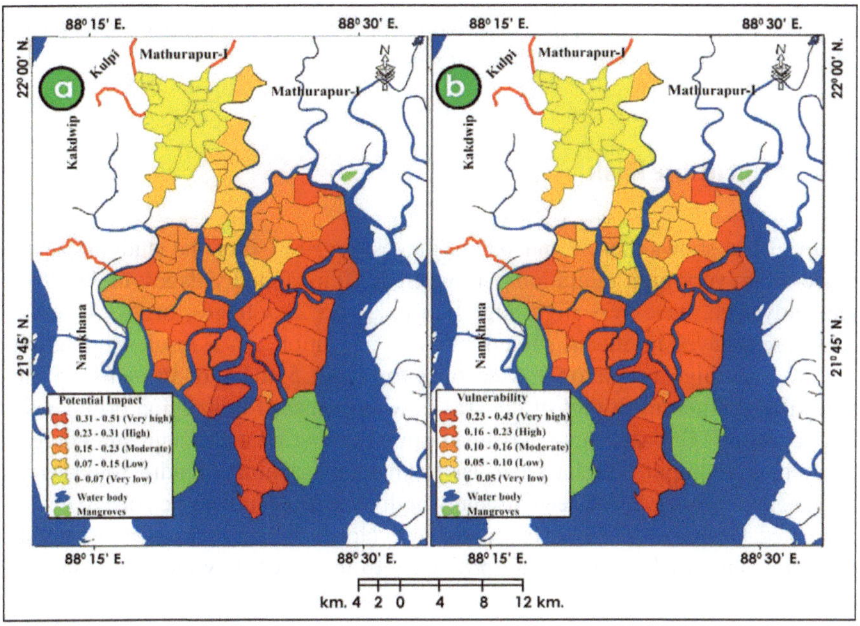

Fig. 3.12 Spatial distribution of **a** potential impact (exposure and sensitivity) and **b** vulnerability (based on exposure, sensitivity and adaptive capacity)

3.4.3 Vulnerability-Integrated Adaptive Capacity

The produced cyclone vulnerability map with adaptive capacity highlights regions where low mitigation capability affects the vulnerability assessment and displays conclusions that differ from those of the vulnerability assessment without adaptive capacity (Fig. 3.12b). Due to the integration of adaptive capacity with the potential impact, it was discovered that while adaptive capacity was not significantly impacted in places with high cyclone risk, it was significantly impacted in areas with moderate to low hazard, helping to mitigate the negative effects of the cyclone and causing moderate to low vulnerable regions to emerge.

The spatial distribution of vulnerability showed that 10.34% and 24.14% of the total area fall into the very high and high classes, respectively (Table 3.3). This is a reduction of 10% in vulnerability from the scenario of vulnerability without adaptive capability. Moreover, 34.48% of villages were in less vulnerable areas; but, after the integration of adaptive capacity, this number has decreased to less than 34%. As a result, 10% of the region has been transformed from vulnerable classes into less vulnerable sections. Kshetra Mohanpur, Chhoto Banashyamnagar, Gangapur, Gobindapur Abad, Paschim Sripatinagar, and lastly Sibnagar are the villages that have transitioned from a very high to highly vulnerable class. Dakshin Gobindapur, Heramba Gopalnagar, Kamdebpur, Maheshpur, Radhakrishnapur Nagar, and last but not least Sagar Madhabpur are moderate to low vulnerable zone transformed villages.

After the integration of adaptive capacity, Purna Chandrapur village moves up to the very low susceptible class from the low class.

3.5 Migration Risk

Figure 3.13a shows that 52 inhabited villages, or 59.77% of the study area, have moderate to very-high migration risk, whereas 10 inhabited villages, or 11.49% of the study area, have very high migration risk, and 21 villages, or 24.14% of the study area, have significant susceptibility to migration. As a result, regions with high to very high migration risk account for about 53.63% of the villages' land area. The southernmost part of the area is where most of this region originates. A large portion of Gobardhanpur, Krishnadaspur, Buraburir Tat, Dakshin Surendraganj, Sitarampur, Indrapur, Rakshaskhali, and Rakhalpur, Paschim Surendra Nagar, as well as Gangapur mouza, exhibited extremely high migration vulnerable zone to cyclone generated. Cropland is heavily inundated by cyclones, which are caused by a number of significant variables, including proximity to the coast, low elevation, and high cyclone tracks.

Other socio-economic indicators include having a kucha wall and roof, poor literacy, fewer people with more than Rs. 10,000 monthly income, fewer permanent structures, and reliance on a single crop (low crop diversity). This area is also where the majority of cop land is damaged during cyclones, leading to the loss of

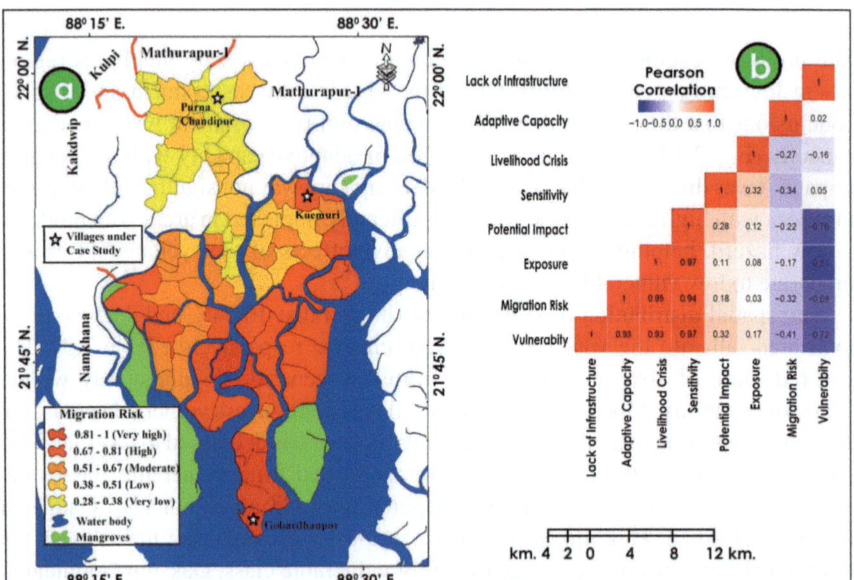

Fig. 3.13 Spatial distribution of **a** migration risk and **b** correlation matrix

crops. Taranagar, Malaya, Piprakhali, Digambarpur, Purna Chandrapur, Jagindrapur, Ramdebpur, Dakshin Raypur, Biswanathpur, Dakshin Durgapur, Nilerat, and Gurudaspur are all from the northern part of the area, which is away from the coastline, has no or fewer cyclone tracks, and has well-developed socio-economic conditions, making it a low migration risk-prone area. The table shows that the area of the study area with a very low migration risk is covered by 13.79% of the area (which includes 12 villages), and the low migration risk zone by roughly 26.44% (23 inhabited villages) (Table 3.4).

3.5.1 Case Study on the Migration Risk Model

Any model that has ever been created based on a theoretical framework or assumption variables must be validated with a real database in order to be accepted under all conditions. Remind everyone once more that our study, which employed proxy factors instead of real migration data to identify a probable risk zone owing to cyclone-induced migration, was completed. So far, a case study has allowed us to evaluate our model using actual data.

Our model's classification leads to the initial selection of three villages. Our migration risk mapping model was divided into five categories, and we have selected three villages from each category: one is in the low migration risk zone, one is in the moderate migration risk zone, and the third village is in the very high migration risk category. Geographically speaking, these villages are descended from Gobardhanpur village, which is considered a coastal village because it is located on the coast, at the first point village of the Pathar Pratima toward the coast. The second Kuemuri village, which is considered a riverine village and is located on the river's edge, and the final Kuemuri village, which is considered an inland village and has no direct access to the sea or river, are both treated as villages. As we've concluded from the analysis of our study, the riverine village (Kuemuri) is comparatively less risky than the coastal village, with the closest hamlet to the coast being in the most vulnerable position. Inland communities are less vulnerable than coastal or riverine settlements because they are less affected by cyclones and less susceptible to river bank erosion or embankment breaches (Fig. 3.14).

Following the selection of the villages, a questionnaire survey, as well as open conversion one-to-one and focus group discussions (FGD), were conducted in order to gather the data from the field survey. By using both formal and informal conversation, the opinions regarding the impact of migration (qualitative) and migration rate (quantitative) were both collected. With the interviewer's permission, the conversation was also recorded as video evidence to help support the findings. The field was finished before the "Mocha" cyclone made landfall on May 9th and 10th, 2023. Though the Indian Sundarbans was not affected by Cyclone "*Mocha*."

There were two different sorts of questions in the questionnaire that had to do with human migration, such as how people saw the influence of many hazards, on their decision to migrate. A 3-point Likert scale with the options "high impact on

Table 3.4 Correlation matrix between migration risk and with other exposure and adaptive capacity

	Hazard	Sensitivity	Adaptive capacity	Livelihood crisis	Lack of infrastructure index	Potential impact	Vulnerability	Migration risk
Hazard	1.00							
Sensitivity	0.11	1.00						
Adaptive capacity	−0.17	−0.34	1.00					
Livelihood crisis	0.93	0.09	−0.22	1.00				
Lack of infrastructure index	−0.81	0.05	0.01	−0.72	1.00			
Potential impact	0.97	0.28	−0.21	0.90	−0.76	1.00		
Vulnerability	0.93	0.31	−0.41	0.87	−0.72	0.97	1.00	
Migration risk	0.90	0.20	−0.33	0.97	−0.58	0.90	0.90	1.00

Source Computed by the Researcher

Fig. 3.14 Being an island community by a river, Purba Sripatinagar is extremely vulnerable to environmental hazards such cyclonic storms, tidal flooding, and riverbank erosion. Its economic backwardness has been greatly exacerbated by its remote location and frequent vulnerability to natural disasters; the majority of its citizens live in poverty and lack access to reliable sources of income and basic infrastructure. *Source* By authors on 07-07-2023

migration," "moderate impact and low impact" was used to collect respondents' opinions on the topic of "*multi-hazards induced seasonal*" *migration to nearest cities or other places.* Dummy variables were employed to quantify the data; three were utilized for strong impact, two for moderate impact, and one for low impact. Fifty families from each village were polled, and we attempted to determine the impact of the three opinions (high impact [3], moderate [2], and low impact [1]) on each of the 50 households. According to Table 3.5, due to the high impact of migration on Gobardhanpur, where this village is located, 12% of 50 households there are concerned. Whereas at Kuemuri, 28% of the households said the high impact of migration, whereas at Purna in the inland community of Chandipur, 16% of households are concerned about the significant effects of migration. On the other hand, Kuemuri and 52% and 36%, respectively, have higher and lower perceptions of the impact of migration. Moderate to low impact perception is ranked last at Purna Chandipur with 46% and 38%, respectively.

On the other hand, households that had moved and those that had not were used to compute the actual migration rate. According to our proposed model, Gobardhanpur has a relatively high migration risk zone, with 28% of households migrating there compared to 24% in Kuemuri and 12% in Purna Chandipur (Table 3.6). Therefore, the migration data at the unit level from the three villages are 14, 12, and 11, respectively. Therefore, even though the perception of migration was not consistent with the model, the actual rate of migration has been consistent with our migration risk model, with the coastal village of Gobardhanpur having the highest migration risk of the three

Table 3.5 Perception of migration

Class	Gobardhanpur (No. of HH)	Buraburir tat (share)	Kuemuri (No. of HH)	Kuemuri (share)	Purna Chandipur (No. of HH)	Purna Chandipur (share)
High impact	6	12	14	28	8	16
Moderate	26	52	27	54	23	46
Low impact	18	36	9	18	19	38

Source Computed by the Researcher based on the Primary Survey

villages, while the riverine village has a migration rate that is lower than that of a coastal village but higher than that of an inland village.

There was an open dialogue with a Gobardhanpur local. She had been a widow for 30 years and was elderly. She claimed that "nature (the water) has changed so much up until this point". Their home was distant from the coast where they were currently. The coastal embankment or line was rebuilt inland two to three times. Farmland has vanished into the sea. The household relies on seaside fishing and farming, when it was specifically stated that you didn't want to live and work outside for a better standard of living. "Her only son, having with son's wife, and grandson and granddaughter," was her response. So, her family just consists of one person. Thus, it is challenging to tell that her son has moved outside. Despite the fact that this region is highly exposed to cyclone dangers and that residents should relocate for a better quality of life, many choose not to do so because of a lack of resources and a lack of social capital. The family is now able to arrive as a "*Trapped Population.*" At Kuemuri, another interview was conducted. The interviewer is between the ages of 30 and 35. He worked in migration. He moved in the years 2015 and 2016, first to Bangalore and then to Kolkata, where he worked for an IT company before returning to his hometown in 2019. As commodities are carried from one location to another inside the country, he is now engaged in fishing, cultivation, and small enterprises.

Table 3.6 Migration rate village wise

Villages	Migration (share)	Migration risk
Gobardhanpur	28	Very high
Kuemuri	24	High
Purna Chandipur	22	Low

Source Computed by the Researcher based on the Primary Survey

3.6 Discussion

The livelihood of the coastal community is thought to be weakened by susceptibility and the degree of cyclonic exposure (Rakib et al. 2019). Four strong cyclones were observed during the past 5 years, and the loss of household infrastructure was disproportionately large to family income (Mondal et al. 2021). According to Mondal et al. (2021), the disruption of social structure and economic loss are two of the main effects of tropical cyclones. The role of coping capacity covers adaptive capability, resource availability, and social capital, which are more robust during post-disaster times and can make people more vulnerable and push them to leave the area (Rakib et al. 2019). The majority of people survived the losses and harm caused by natural resources by fishing in the river, taking out a loan, receiving aid, and participating in aid (Mallick and Siddiqui 2015). Cyclone strikes, property losses, saline water infiltration, destruction of houses and roads, and loss of employment prospects were the main causes of poverty in the study region (Fig. 3.15). According to our findings, the majority of the households in Pathar Pratima (almost 70%) are fragile, temporary buildings that are extremely vulnerable to natural calamities. With an average elevation of almost 2 m, a major portion of Pathar Pratima CD Block is still below the high tide line (Mondal et al. 2021). Climate induced extreme events have a major impact on those who work in basic industries, particularly farmers and fishermen (Mistri and Das 2014; Haldar and Debnath 2014). The average population percentage for backward communities is over 25%. Results indicate that more than 70% of households in Ramdebpur Mouza are considered to be deprived, where the large proportions of backward communities are found (Socio-economic Caste Census 2011). Due to their low living standards, the scheduled caste and scheduled tribe populations are, in comparison to other communities, found most susceptible (Mistri and Das 2014). The pilot survey demonstrated that the southern and south-eastern regions of Pathar Pratima are more vulnerable than the rest of the region, as these backward communities settled within the proximity to sea, which puts them at greater risk from natural disasters. Yet, an upward trend in sea surface temperature (SST) of almost 0.5 °C per decade aids in the development of stronger and more deadly cyclones over the Bay of Bengal (Hazra et al. 2002; Singh 2007).

It is considered to be vulnerable if the area is both highly exposed to environmental hazards and extremely sensitive to their impacts, and has a limited capacity to deal with those hazards. On the other hand, if a system has both a strong adaptive capacity and low exposure to environmental hazards, it will be less susceptible to vulnerability (Smit and Wandel 2006; Fellmann 2012; Roy et al. 2021). Our ability to assess vulnerability is aided by the combined impacts of exposure, sensitivity, and adaptive capability. The study established that there has been too much influence of vulnerability on climatic migration risk, as we see the r^2 value is 0.82 (Fig. 3.16a). The cumulative effects of exposure, sensitivity, and adaptive capacity assist us in determining vulnerability level. The potential impact is individually correlated with Exposure ($r = 0.97$) and sensitivity ($r = 0.28$) as positive, but the combined effects of these two components are considerably increased. As a result, aggregate values'

Fig. 3.15 Silent Struggles and echoes of Pathos: as sons pursue livelihood opportunities in the Southern States of India, a partially blind elder navigates loneliness. *Source* By authors on 10-01-2024

potential impact combination of exposure and sensitivity are more strongly correlated ($r = 0.97$) with vulnerability rather than individual values.

Livelihood crisis has a positive correlation with a value of 87% but infrastructure availability has a negative correlation (0.72). The components of exposure, sensitivity, and livelihood crises are positively correlated with migration risk, with covariances of 0.90, 0.20, and 0.58, respectively. But Migration risk was negatively correlated (Table 3.4) with adaptive capacity ($r = -0.33$) and infrastructure availability ($r = 0.58$). The three key factors of vulnerability—vulnerability as measured by the average of Exposure, Sensitivity, and Adaptive Capacity—livelihood crisis, and infrastructure accessibility—were used to assess migration risk. Here, the exposure and agricultural vulnerability point to a positive correlation between migration risk and availability of infrastructure, while a negative correlation exists between the two. In order to achieve unidirectional infrastructure availability, the entire component is being done here. With respective covariances of 0.90, 0.20, and 0.58, the components of exposure, sensitivity, and livelihood crises have a positive relationship with migration risk (Fig. 3.16b).

In the past three decades, every mouza has been badly impacted by the average of three large cyclones that hit the entire region. Specifically for cyclone Aila in May 2009, a vast area in Pathar Pratima has been found inundated (Fig. 3.16c), and this scenario was also recitative during Amphan (May 2020) and Yaas cyclone (May

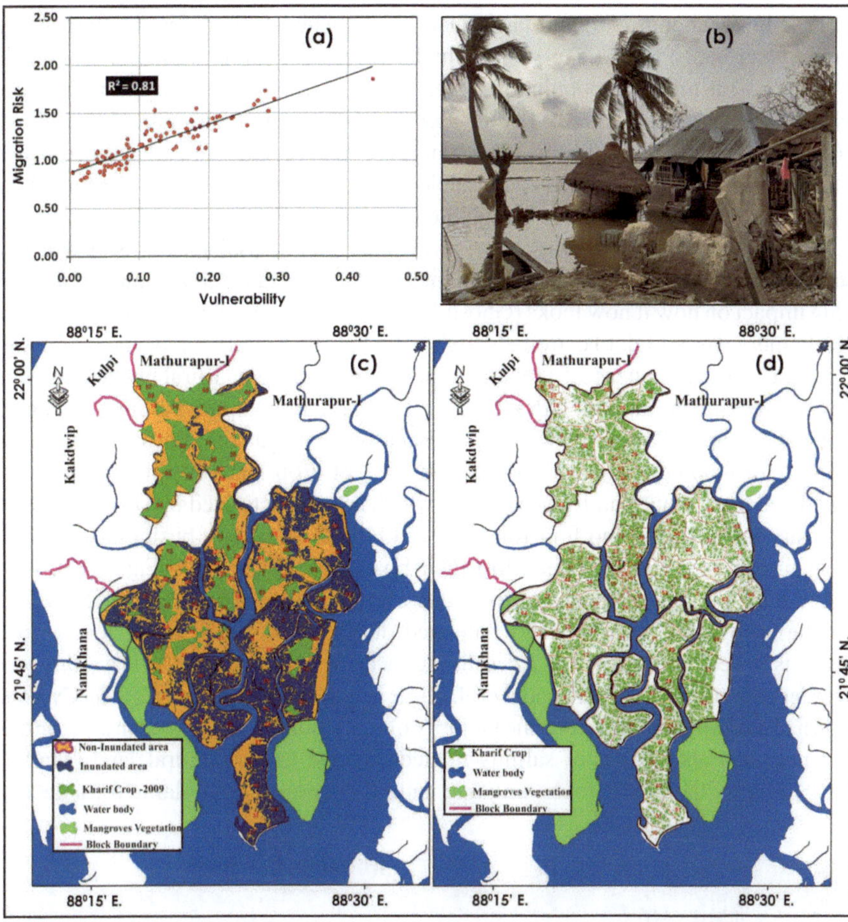

Fig. 3.16 The scenarios of cyclone and the condition of livelihood crisis: **a** the relationship between vulnerability and migration risk, **b** the real situation is faced after the cyclone in a coastal village of Pathar Pratima, **c** the status of Kharif crop (Oct–Nov, 2009) in the cyclone year Aila (2009, May), and **d** the status of Kharif crop (Oct–Nov, 2017) during a non-cyclone year (2017)

2021). The southern villages Gobardhanpur, Sitarampur, Dakshin Surendranagar, Indrapur, and Buraburir Tat have likely had more than four or close to five (4.50) cyclones in the past 30 years, making them the places with the highest cyclonic exposure. The average inundation area of the last three major cyclones, *Bulbul* (2019), *Amphan* (2020), and *Yaas* (2021), was used here as well to create an agricultural inundation map. It was found that 22.18% of the cropland in every mouza was inundated, with the most affected villages (Table 3.2) being Gobardhanpur (50.21%), Sitarampur (45.52%), Buraburir (45.45%), Krishnadaspur (44.76%), and Rakshaskhali (42.56%). The maximum population density was in Banashyamnagar, Bardapur, Bhagbatpur, Abad Gangadharpur, Dakshin Lakshmi Narayanpur, followed by Dakshin Mahendrapur, and lastly Dakshin Madhusudanpur. These villages also have kucha walls and kucha roofs, which make them more sensitive. Compared to places with minimal populations, these locations are more susceptible to tropical cyclones (Ali et al. 2020). For a better understanding of agricultural livelihood, Fig. 3.16d shows a normal year, 2017, where we can see that more and less crop area was cultivated, but these were not possible to cultivate in 2009 due to Cyclone Aila, which damaged the crop area through the inundation of saline water.

The sensitivity factors intensify the level of hazard risk (Sahana and Sajjad 2019). It shows how the system reacts to environmental dangers' detrimental impacts and has a big impact on how it now looks (Ghosh and Mistri 2021). In the south and east of the block, there was a high to extremely high potential for migration risk. These regions are a part of the mouzas of Gobardhanpur, Krishnadaspur, Buraburirtat, Dakshin Surendraganj, Sitarampur, Indrapur, Rakshaskhali, Rakhalpur, Paschim Surendra Nagar, Gangapur, Paschim Dwarakapur, Banashyam Nagar, and Purba Sripatinagar, along with Gobindapur Abad. More than 35% of crop lands in these regions are inundated by cyclones and storm surges, which are characterized by strong cyclonic intensity (affected around four times). Due to the storm surge, businesspeople lost their groceries, while fishermen lost their boats and fishing nets (Fig. 3.17). As a result, they lost their source of income and eventually fled (Mehedi et al. 2010).

The crop fields were severely salinized during Aila, and it took close to two sessions to restore fertility. As a result, the locals were compelled to leave this area in order to continue living (Mondal et al. 2021; Kayal and Chowdhury 2022). Cyclone Aila had a significant impact on the majority of individuals who worked in agriculture, and migrants also spoke of salinity-related damage to agricultural land (Mallick and Siddiqui 2015). Moreover, ongoing natural disasters are to blame for a decline

Fig. 3.17 Fishing cultivation is the second most important livelihood activity. *Source* By authors on 18-11-2023

in human resilience, which ultimately compelled people to migrate. These natural disasters have a negative impact on income (Siddiqui 2014). With the potential for beneficial adaptation co-benefits, migration was primarily adopted as a response to risk and uncertainty (Maharjan et al. 2020). According to Martin et al. (2014), better wages are the main reason why villagers in communities that are increasingly subject to cyclonic pressures and shocks are diversifying their native livelihood strategies by relocating. Migration evolves into a beneficial adaptation strategy, and it is also recognized by society (Martin et al. 2014). At risk exposure levels, there is a tendency for greater socio-economic inequality at the community level, which suggests that those who are most at risk for moving in order to protect their way of life after an environmental hazard are those who are most vulnerable (Mallick and Vogt 2014).

3.7 Conclusions

Globally, migrations have played an essential role in the history of humanity, and some of them are caused by climate hazards and environmental change. The International Organization for Migration (IOM) defines environmental migration as human movement caused by abrupt or gradual environmental changes that negatively affect people's quality of life and force them to leave their usual residences either freely or involuntarily within their own nations or abroad, either temporarily or permanently (IOM 2011). Some research (such as IOM 2018; Jha et al. 2018; Luetz and Merson 2020; Twinomuhangi et al. 2023) views migration as a combined cause of climate change and adaptation strategy, while others contend that environmental and climate changes are multiplier stressors that trigger migration (Scheffran et al. 2012; UNFCCCC 2015). The study also showed that cyclonic hazards to human mobility are the most important element in this area, and migratory decisions made in this region are more of a coping technique than a component that significantly influences adaptability. Hence, academics have proven via extensive empirical research that there are still many unanswered concerns regarding climate migration. In order to choose the survey area for an empirical study, this study provides guidance to academics, government agencies, and non-governmental groups planning to manage this issue.

References

Adger WN, Agrawala S, Mirza MMQ, Conde C, O'Brien K, Pulhin J, Pulwarty B, Takahashi K (2007) Assessment of adaptation practices, options, constraints and capacity. Clim Chang 200

Ali SA, Khatun R, Ahmad A, Ahmad SN (2020) Assessment of cyclone vulnerability, hazard evaluation and mitigation capacity for analyzing cyclone risk using GIS technique: a study on Sundarban biosphere reserve, India. Earth Syst Environ 4:71–92

Alizadeh M, Alizadeh E, AsadollahpourKotenaee S, Shahabi H, Beiranvand Pour A, Panahi M et al (2018) Social vulnerability assessment using artificial neural network (ANN) model for earthquake hazard in Tabriz city, Iran. Sustainability 10(10):3376

Baldwin A, Methmann C, Rothe D (2014) Securitizing 'climate refugees': the futurology of climate-induced migration. Crit. Stud. Secur. 2(2):121–130

Bhardwaj P, Pattanaik DR, Singh O (2019) Tropical cyclone activity over Bay of Bengal in relation to El Niño-Southern Oscillation. Int J Climatol 39(14):5452–5469

Black R, Adger WN, Arnell NW, Dercon S, Geddes A, Thomas D (2011) The effect of environmental change on human migration. Glob Environ Chang 21:S3–S11

Brown S, Nicholls RJ (2015) Subsidence and human influences in mega deltas: the case of the Ganges-Brahmaputra-Meghna. Sci Total Environ 527–528:362–374. https://doi.org/10.1016/j.scitotenv.2015.04.124

Chand BK, Trivedi RK, Biswas A, Dubey SK, Beg MM (2012) Study on impact of saline water inundation on freshwater aquaculture in Sundarban using risk analysis tools. Explor Anim Med Res 2(2):170–178

Council–NRC, N. R. Internal Displacement Monitoring Center-IDMC (2014) Global overview 2014. In: People internally displaced by conflict and violence. https://reliefweb.int/report/india/asia-and-pacific-weekly-regional-humanitarian-snapshot-19-25-may-2020

Das S, Ghosh A, Hazra S, Ghosh T, de Campos RS, Samanta S (2020) Linking IPCC AR4 and AR5 frameworks for assessing vulnerability and risk to climate change in the Indian Bengal delta. Prog Disaster Sci 7:100110

Dubey SK, Trivedi RK, Chand BK, Mandal B, Rout SK (2017) Farmers' perceptions of climate change, impacts on freshwater aquaculture and adaptation strategies in climatic change hotspots: a case of the Indian Sundarban delta. Environ Dev 21:38–51

Fellmann T (2012). The assessment of climate change-related vulnerability in the agricultural sector: reviewing conceptual frameworks. In: Building resilience for adaptation to climate change in the agriculture sector, vol 23, p 37

Ghosh S, Mistri B (2020) Geo-historical appraisal of embankment breaching and its management on active tidal land of Sundarban: a case study in Gosaba Island, South 24 Parganas, West Bengal. Space Cult, India 7(4):166–180

Ghosh S, Mistri B (2021) Assessing coastal vulnerability to environmental hazards of Indian Sundarban delta using multi-criteria decision-making approaches. Ocean Coast Manag 209:105641

Ghosh S, Mistri B (2022) Analyzing the multi-hazard coastal vulnerability of Matla-Bidya inter-estuarine area of Indian Sundarbans using analytical hierarchy process and geospatial techniques. Estuar Coast Shelf Sci 279:108144

Giri S, Samanta S, Mondal PP, Basu O, Khorat S, Chanda A, Hazra S (2022) A geospatial assessment of growth pattern of aquaculture in the Indian Sundarbans biosphere reserve. Environ Dev Sustain 24(3):4203–4225

Gray CL, Mueller V (2012) Natural disasters and population mobility in Bangladesh. Proc Natl Acad Sci 109(16):6000–6005

Grothmann T, Patt A (2005) Adaptive capacity and human cognition: the process of individual adaptation to climate change. Glob Environ Chang 15(3):199–213

Haldar A, Debnath A (2014) Assessment of climate induced soil salinity conditions of Gosaba Island, West Bengal and its influence on local livelihood. In: Climate change and biodiversity: proceedings of IGU Rohtak conference, vol 1. Springer Japan, pp 27–44

Hazra S, Ghosh T, DasGupta R, Sen G (2002) Sea level and associated changes in the Sundarbans. Sci Cult 68(9/12):309–321

Hoque MAA, Pradhan B, Ahmed N, Ahmed B, Alamri AM (2021) Cyclone vulnerability assessment of the western coast of Bangladesh. Geomat Nat Haz Risk 12(1):198–221

IOM (2011) Glossary on migration, 2nd edn. International Organisation for Migration, Geneva, Switzerland. https://publications.iom.int/system/files/pdf/iml25_1.pdf

IOM (2018) Migration, environment and climate change: definitional issues. International Organisation for Migration, Geneva, Switzerland. https://www.iom.int/migration-and-climate-change
IPCC (2014a) Climate change (2014). Synthesis report. In: Contribution of working groups I, II and III to the fifth assessment report of the intergovernmental panel on climate change
IPCC (2014b) Climate change 2014. Impacts, adaptation and vulnerability, fifth assessment report. In: Intergovernmental panel on climate change. Cambridge University Press, Cambridge, UK
Jha CK, Gupta V, Chattopadhyay U, Amarayil Sreeraman B (2018) Migration as adaptation strategy to cope with climate change: a study of farmers' migration in rural India. Int J Clim Chang Strat Manag 10(1):121–141
Karimi A, Abdollahi S, Ostad-Ali-Askari K, Eslamian S, Singh VP (2021) Predicting fire hazard areas using vegetation indexes, case study: forests of Golestan Province, Iran. J Geogr Cartogr 4(1):1–6
Kayal P, Chowdhury IR (2022) Impact of frequent cyclonic storms on land use and land cover changes along the coastal areas of Indian Sundarban. India II: climate change impacts, mitigation and adaptation in developing countries. Springer International Publishing, Cham, pp 33–71
Kumar PD, Paul YS, Muraleedharan KR, Murty VSN, Preenu PN (2016) Comparison of long-term variability of sea surface temperature in the Arabian Sea and Bay of Bengal. Reg Stud Mar Sci 3:67–75
Le TVH, Nguyen HN, Wolanski E, Tran TC, Haruyama S (2007) The combined impact on the flooding in Vietnam's Mekong River delta of local man-made structures, sea level rise, and dams upstream in the river catchment. Estuar Coast Shelf Sci 71(1–2):110–116
Lin N, Emanuel K, Oppenheimer M, Vanmarcke E (2012) Hurricane surge and global warming: a physically-based risk assessment. Nat Clim Chang 2:462
Luetz JM, Merson J (2020) Climate change and human migration as adaptation: conceptual and practical challenges and opportunities. Clim Action 120–132
Maharjan A, de Campos RS, Singh C, Das S, Srinivas A, Bhuiyan MRA et al (2020) Migration and household adaptation in climate-sensitive hotspots in South Asia. Curr Clim Chang Rep 6:1–16
Mallick B, Siddiqui T (2015) Disaster-induced migration and adaptation discourse in Bangladesh. In: Environmental change, adaptation and migration: bringing in the region, pp 164–185
Mallick B, Vogt J (2014) Population displacement after cyclone and its consequences: Empirical evidence from coastal Bangladesh. Nat Hazards 73:191–212
Mandal UK, Nayak DB, Samui ARPAN, Jana AK, Mullick SOURAV, Lama TD et al (2018) Trend of sea-level-rise in West Bengal coast
Martin M, Billah M, Siddiqui T, Abrar C, Black R, Kniveton D (2014) Climate-related migration in rural Bangladesh: a behavioural model. Popul Environ 36:85–110
Mehedi H, Nag AK, Farhana S (2010) Climate induced displacement. In: Case study of cyclone Aila in the southwest coastal region of Bangladesh
Mistri A, Das B (2014) Assets disparities among social groups: a cross sectional analysis of census 2011. Soc Chang 44(1):1–20
Mondal M, Haldar S, Biswas A, Mandal S, Bhattacharya S, Paul S (2021) Modeling cyclone-induced multi-hazard risk assessment using analytical hierarchical processing and GIS for coastal West Bengal, India. Reg Stud Mar Sci 44:101779
Mondal M, Biswas A, Haldar S, Mandal S, Bhattacharya S, Paul S (2022) Spatio-temporal behaviours of tropical cyclones over the bay of Bengal Basin in last five decades. Trop Cyclone Res Rev 11(1):1–15
New M, Hewitson B, Stephenson DB, Tsiga A, Kruger A, Manhique A et al (2006) Evidence of trends in daily climate extremes over southern and west Africa. J Geophys Res: Atmos 111(D14)
Pal S, Paul S (2021) Stability consistency and trend mapping of seasonally inundated wetlands in Moribund deltaic part of India. Environ Dev Sustain 23(9):12925–12953
Perch-Nielsen SL, Bättig MB, Imboden D (2008) Exploring the link between climate change and migration. Clim Chang 91:375–393

Pourghasemi HR, Gayen A, Edalat M, Zarafshar M, Tiefenbacher JP (2020a) Is multi-hazard mapping effective in assessing natural hazards and integrated watershed management? Geosci Front 11(4):1203–1217
Pourghasemi HR, Kariminejad N, Amiri M, Edalat M, Zarafshar M, Blaschke T, Cerda A (2020b) Assessing and mapping multi-hazard risk susceptibility using a machine learning technique. Sci Rep 10(1):3203
Rahman M, Ghosh T, Salehin M, Ghosh A, Haque A, Hossain MA, Das S, Hazra S, Islam N, Sarker MH, Nicholls RJ (2020) Ganges-Brahmaputra-Meghna delta, Bangladesh and India: a transnational mega-delta. In: Deltas in the Anthropocene. Palgrave Macmillan, Cham, pp 23–51
Rakib MA, Sasaki J, Matsuda H, Fukunaga M (2019) Severe salinity contamination in drinking water and associated human health hazards increase migration risk in the southwestern coastal part of Bangladesh. J Environ Manage 240:238–248
Roy S, Bose A, Chowdhury IR (2021) Flood risk assessment using geospatial data and multi-criteria decision approach: a study from historically active flood-prone region of Himalayan foothill, India. Arab J Geosci 14(11):999
Sahana M, Sajjad H (2019) Vulnerability to storm surge flood using remote sensing and GIS techniques: a study on Sundarban biosphere reserve, India. Remote Sens Appl Soc Environ 13:106–120. https://doi.org/10.1016/j.rsase.2018.10.008
Sahoo B, Bhaskaran PK (2016) Assessment on historical cyclone tracks in the Bay of Bengal, east coast of India. Int J Climatol 36(1):95–109
Scheffran J, Marmer E, Sow P (2012) Migration as a contribution to resilience and innovation in climate adaptation: social networks and co-development in Northwest Africa. Appl Geogr 33:119–127
Schoonover R, Cavallo C, Caltabiano I (2021) The security threat that binds us: the unraveling of ecological and natural security and what the United States can do about it. The Converging Risks Lab, Institute of the Council on Strategic Risks, Washington, DC
Shultz JM, Russell J, Espinel Z (2005) Epidemiology of tropical cyclones: the dynamics of disaster, disease, and development. Epidemiol Rev 27(1):21–35
Siddiqui R (2014) Patterns and factors of natural hazard induced out-migration from Meghna Estuarine Islands of Bangladesh. GeoScape 8(1):17–31
Singh OP (2007) Long-term trends in the frequency of severe cyclones of Bay of Bengal: observations and simulations. Mausam 58(1):59–66
Smit B, Wandel J (2006) Adaptation, adaptive capacity and vulnerability. Glob Environ Chang 16(3):282–292
Su YY (2014) The one billion ‘climate refugees’ that never was: INGOs and the human rights perspective to climate change-induced displacement. Oxford Monitor of Forced Migration
Sullivan C (2002) Calculating a water poverty index. World Dev 30(7):1195–1210
Twinomuhangi R, Sseviiri H, Kato AM (2023) Contextualising environmental and climate change migration in Uganda. Local Environ 1–22
Uddin MJ, Li Y, Cheung KK, Nasrin ZM, Wang H, Wang L, Gao Z (2019) Rainfall contribution of tropical cyclones in the Bay of Bengal between 1998 and 2016 using TRMM satellite data. Atmosphere 10(11):699
UNDP (United Nations Development Programme, Bureau for Crisis Prevention) (2004) Reducing disaster risk: a challenge for development—a global report. United Nations
UNDP (2007) Human development reports 2007/8. In: Fighting climate change: human solidarity in a divided world
UNFCCC (2015) Human mobility in the context of climate change UNFCCC-PARIS COP-21. Recommendations from the advisory group on climate change and human mobility. UNHCR. 2015. https://www.unhcr.org/protection/environment/565b21bd9/human-mobility-context-climate-change-unfccc-paris-cop-21-recommendations.html
Wahiduzzaman M, Oliver EC, Wotherspoon SJ, Holbrook NJ (2017) A climatological model of North Indian Ocean tropical cyclone genesis, tracks and landfall. Clim Dyn 49:2585–2603

Chapter 4
Pattern and Socio-economic Condition of Rural Migrants from Coastal Sundarbans

Abstract The practice of migration from Sundarbans (India) has become a normal phenomenon, mainly after cyclone Aila (2009), and it also again rises due to the cyclonic damage of cyclone Amphan (2020). Amphan made the situation most vulnerable and compelled the coastal people to migrate, though it was on a temporary basis as migrant workers. This migration could be called climatic migration. The chapter's main objective is to understand the migration pattern and the socio-economic condition of both migrant and non-migrant households. The study was executed through empirical data collected from 750 households by random sampling and categorized as migrant households, meaning a minimum of one person migrated from their family, and no one migrated from a family treated as a non-migrant household. The finding revealed that the majority of the migrants are young and belong to lower backward communities socially and also economically, like the SC and Muslim communities, having low landholding capacity, interpreting the low resilient people have more chances to migrate to another place as migrants work. Most people migrate from the Sundarbans to the southern states of India, like Tamil Nadu, Kerala, and Karnataka, where job opportunities and labor wages are also high compared to their native place. The majority of the migrant laborers work in the construction sector, and they also come back to their homes twice a year. The study also found that their income is raised after migration. They initially construct concrete houses through their remittance, become more resilient, and enhance their quality of life compared to non-migrant households. This migration, as revealed by the study, is not just a response to the adverse effects of the cyclones, but also an adaptation strategy for the coastal dwellers to cope with the changing climate.

Keywords Migration · Sundarbans · Coastal · Inland · Landholding capacity · Remittance · Non-migrant

P. Mandal and S. Paul, *Climatic Vulnerability and Migration: Evidence from Indian Sundarbans*, Springer Geography, https://doi.org/10.1007/978-3-032-05858-4_4

4.1 Introduction

Discussions on "climate migration" increasingly acknowledge the need for studies on actual behavior and a more intricate probabilistic approach. This approach involves analyzing climate stressors alongside other socio-economic and political drivers (Hunter et al. 2015). The International Organization for Migration defines environmental migration as human mobility prompted by sudden or gradual environmental changes that negatively affect people's living conditions, compelling them to leave their usual homes, either voluntarily or involuntarily, within their own countries or abroad, for a temporary or permanent period (Black et al. 2013). In its fifth assessment report, the Intergovernmental Panel on Climate Change (IPCC) suggests that climate change is a significant factor driving migration, particularly climatic and environmentally induced migration within nations (Schoonover et al. 2021). Furthermore, Schoonover et al. (2021) emphasize that environmental changes interact with climate change, posing food, economic, and livelihood security risks, thereby fueling migration. Given the critical role of the environment, especially the climate, in sustaining livelihoods, its degradation jeopardizes people's socio-economic well-being, prompting migration (Mondal et al. 2023). Environmental migration ultimately arises from the sudden onset of extreme climate events and catastrophes such as storms, floods, landslides, and earthquakes, causing displacement and relocation. It also stems from gradual onset processes like rising temperatures, land degradation, and the long-term effects of dry spells and drought on agriculture and ways of life (Twinomuhangi et al. 2023; Mondal et al. 2023).

The intricate relationship between population movement and climate change is embodied in the "climate mobility nexus." The disruption of multiple environmental factors, including temperature, precipitation, sea levels, and extreme weather events, caused by climate change has significant implications for ecosystems, agriculture, water supplies, and infrastructure (Afifi et al. 2016; Dallmann and Millock 2017; Viswanathan and Kumar 2015). Climate-induced or environmental migration is triggered by these changes, leading to the relocation of individuals and groups in pursuit of safer living conditions, either inside their own countries or across international borders. The highlights above highlight the pressing need for all-encompassing policy structures that can effectively tackle the underlying causes of climate change and the many issues displaced people face (Fig. 4.1). These initiatives cover various activities, such as mitigation methods to reduce greenhouse gas emissions, adaptation techniques for changing environmental circumstances, and policies promoting sustainable development and efficient management of migrant flows. Protection of the rights of displaced people by law and measures to build resilience to reduce the probability and effects of displacement events are equally important. By combining these strategies, societies can aim to respond to the intricate relationships between climate change and human mobility in a more resilient and fair way.

The Internal Displacement Monitoring Centre (IDMC) estimates that weather-related events or disasters will likely be the primary cause of 75% of global cases of displacement in 2020 (or roughly 40.5 million people) (IDMC 2021). The variety and

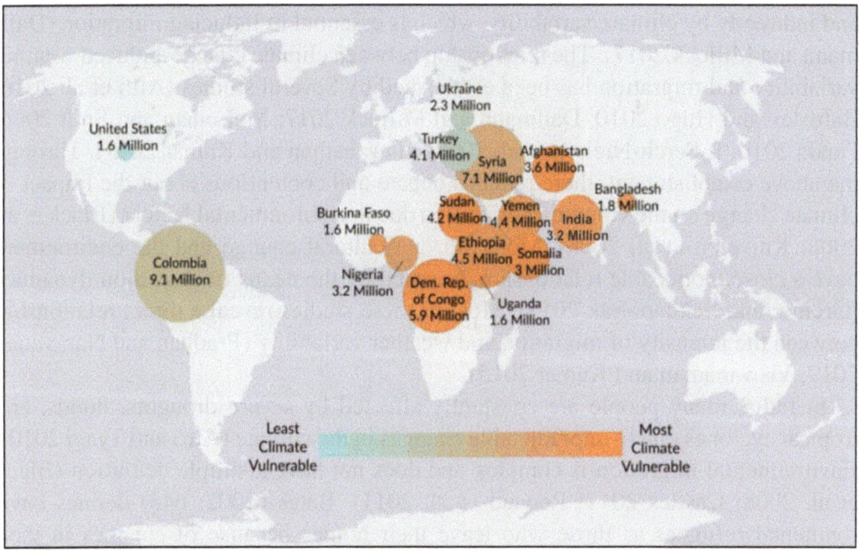

Fig. 4.1 Top humanitarian migrant-hosting countries by population of concern and climate vulnerability level, 2021 (*Source* Migration Policy Institute (MPI) artist rendering based on UNHCR, "Refugee Data Finder," accessed November 15, 2023)

complexity of the relationships between the migration factors make distinguishing between human mobility and environmental reasons challenging, according to the Foresight Report on Migration and Global Environmental Change (Foresight 2011). According to Stapleton et al. (2017), environmental change and climate-related impacts make some locations less habitable, which may cause migration. Critics, however, dispute a lack of data and knowledge regarding the leading causes of environmental migration (Neumann and Hilderink 2015). Understanding how migration and livelihoods are related is necessary to address environmental and climate-induced migrations. Families and individual's dependent on the natural world may change their subsistence methods (Mcleman 2018). Under these circumstances, migration can represent a suitable strategy to cope with and adapt to climate impacts (Foresight 2011; IPCC 2022).

4.2 Cyclones, Livelihood Stress, and Migration

Communities react and migrate for survival, and livelihoods are often determined by Meteorology and climatic changes (Meze-Hausken 2004). The pattern of human migration has been directly or indirectly influenced by climate change and weather variability (McLeman 2013; McLeman et al. 2008; McLeman and Smit 2006). Human health, capital, socio-economic activity, wages, etc., are affected directly and indirectly by climate variability, which is essential in inducing migration (Dallmann and Millock 2017). The relationship between climate change-induced weather variability and migration has been established by Several studies (Afifi et al. 2016; Bardsley and Hugo 2010; Dallmann and Millock 2017; McLeman and Smit 2006; Panda 2010; L Perch-Nielsen et al. 2008; Viswanathan and Kumar 2015). Through the above establishment, there remains debate and contentious about the impact of climate change on the volume of migration due to environmental factors (Black et al. 2008; Kniveton et al. 2008). Moreover, agricultural change and the environment have a close, noticeable relationship, revealed by the nexus of migration dynamics (Greiner and Sakdapolrak 2013). Most of these studies reveal a direct relationship between the intensity of migration and weather variability (Pradhan and Narayanan 2019; Viswanathan and Kumar 2015).

In India, many people are constantly affected by severe droughts, floods, and tropical cyclones due to unpredictable changes in the climate (Attri and Tyagi 2010). Environmental migration is complex and does not have a simple definition (Black et al. 2008; Castles 2017; Renaud et al. 2011). Bates (2002: 648) defines environmental refugees as those who leave their homes because of changes in their natural surroundings. Black et al. (2011) and Adams et al. (2016) emphasize how environmental factors indirectly influence migration by impacting social, economic, political, and demographic aspects. Many rural households face these environmental dangers, pushing them toward migration. However, migration is not solely driven by environmental pressures; it also responds to economic opportunities, leading to disparities between rural and urban areas. Most migrants seek better job prospects to support themselves and their families back home (Ullah 2012; Bernzen et al. 2019), often returning within six months. Research suggests that studies on environmentally induced migration should consider various pressures alongside other migration influencers. Some costs and barriers prevent people from leaving their homes, despite environmental stressors acting as both reasons for and hindrances to migration (Black et al. 2013). While flooding displaces populations temporarily, it does not often result in long-term moves (Gray and Mueller 2012). Young adults migrate within India, South Asia, and beyond for better job prospects (Sengar 2023).

When seasonal migrants move, their social connections often shift. Akhter and Bauer (2014) noted that households with more land are less likely to migrate. Gray and Mueller (2012) and Chen et al. (2017) found that individuals from lower socio-economic backgrounds are less inclined to migrate due to limited resources despite facing greater vulnerability to environmental challenges (Fig. 4.2). Migration researchers like Castles (2017) and Black et al. (2013) emphasize that the cost of moving acts as a significant barrier for those who remain in place, leading to their characterization as "trapped" populations. Some individuals persist in coastal areas despite exposure to or the imminent threat of climatic hazards (Albo et al. 2011; Black et al. 2011; Martin et al. 2014; Mallick and Mallick 2021; Mallick and Schanze 2020; Black and Collyer 2014). Households may diversify their income sources and mitigate risks by relocating household members or relatives, thus enhancing resilience (McLeman and Smit 2006; Gemenne and Blocher 2017). In some communities, migration is a significant method of adapting to environmental stressors, although its

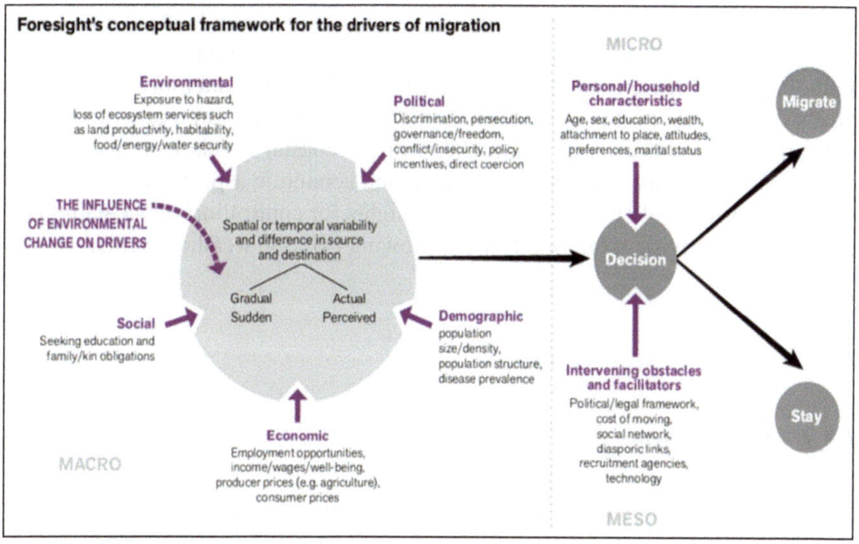

Fig. 4.2 Cyclonic hazards significantly impact migration outcomes, influenced by environmental, political, demographic, economic, and social factors. These factors are interconnected with policies that encourage or impede migration. Cyclonic hazards can exert gradual pressure through prolonged inundation or sudden impacts like tropical cyclones. Individuals may migrate in response to immediate threats or anticipate future risks. The complexity of migration is influenced by these factors and their intertwined nature

suitability varies depending on livelihoods and ecological conditions. When viewed alongside other adaptation strategies, migration appears closely intertwined with and complemented by these activities (Hoffmann et al. 2022).

West Bengal has a rich history of out-migration to various parts of India, a phenomenon extensively explored by Debnath (2020). The Sundarbans, renowned as the most significant global mangrove forest delta and bio-reserve, stands out as one of Earth's most severely affected climatic zones. Following cyclone Aila in 2009, documented by Hajra and Ghosh (2018), climate-induced migration has inflicted significant hardships on the Sundarbans population (Guha and Chandan 2016). Displacement and migration induced by climatic disasters often become the only means of livelihood preservation (Das et al. 2020; Naujoks 2021). In contrast to the Eastern Sundarbans in Bangladesh, the Western part, under Indian jurisdiction, faces even more significant challenges due to multiple climatic, ecological, social, and economic factors (Ghosh and Ghosh 2019). Climatic migration poses a domestic and an international challenge, compelling researchers to explore how environmental factors influence human migration patterns. Despite extensive research on hazards, vulnerability, and multi-hazard risks, issues like cyclone-induced migration still need to be addressed. This empirical study aims to fill this gap by shedding light on the complexities of cyclone-induced migration in a climatic hazard hotspot region, where survival often necessitates migration. Therefore, to effectively manage any weather-related disasters in the future, it is imperative to understand how coastal populations react to cyclones. About migrant and non-migrant households in the southern coastal regions of West Bengal, India, this study attempts to assess the effects of cyclones on their socio-economic status. This research aims to find the motivations underlying rural migration from the coastal Sundarbans and analyze migration trends from demographic, social, and economic aspects and exposure to cyclones. It is possible to gather essential insights for controlling such calamities in the future by studying migration trends following extreme weather events.

4.3 Materials and Methods

4.3.1 Study Area

The study area (Fig. 3.1) encompasses several villages within the Pathar Pratima CD Block in the South 24 Parganas district of West Bengal, India. This region spans approximately 484.47 km^2 and was home to 331,823 people, according to the 2011 Census of India. Pathar Pratima CD Block is bordered by Mathurapur I and Mathurapur II CD Blocks to the north, the Sundarban forests to the east, the Bay of Bengal to the south, and Namkhana, Kakdwip, and Kulpi CD Blocks to the west. It is flanked by the Saptamukhi and Thakuran Rivers in the west and east, respectively, and is characterized by low-lying flat terrain typical of active deltas. Over the past decade, this area has been subject to frequent geomorphologic changes due to its location and has borne the brunt of numerous cyclones. The Human Development Report for South 24 Parganas district, published in 2009, indicated that nearly half (49.13%) of households in the Pathar Pratima CD Block were below the poverty line. Poverty rates were particularly severe in the Sundarban region, with all thirteen CD Blocks reporting poverty rates exceeding 30% and eight CD Blocks having more than 40% of their population classified as below the poverty line. Despite being rich in natural resources, the Sundarban region lags in quality-of-life indicators. A rural household survey conducted in 2005 revealed that 34.11% of households in South 24 Parganas were living below the poverty line, significantly higher than state and national averages. In Pathar Pratima CD Block in 2011, among the total working population, 24.07% were engaged in cultivation, 49.09% worked as agricultural laborers, 3.12% were involved in household industries, and 23.72% were categorized as other workers. This data highlights that approximately 75% of the workforce relied heavily on the physical capital of the region for their livelihoods. The coastal areas of West Bengal, including this region, are highly susceptible to the adverse impacts of cyclones, exacerbating the already challenging living conditions and livelihoods of the residents. Recent cyclones such as Fani, Bulbul, Amphan, and Yaas have further heightened the population's vulnerability to hazards, emphasizing the urgent need for effective disaster management and resilience-building measures.

4.3.2 Criterion for Village Selection

The primary aim of this research is to examine how human migration in the Indian Sundarbans is affected by cyclone-induced hazards. Pathar Pratima, situated in the South 24 Parganas of West Bengal, India, stands out as one of the most vulnerable blocks along the Indian coastal Sundarbans and thus serves as a focal point for this study. However, addressing the issue of human migration, particularly at the village level, poses significant challenges due to a need for comprehensive data on migration patterns and the various factors influencing them, notably environmental factors. This lack of data also complicates efforts to ascertain the true extent of the problem.

Additionally, those who migrate due to environmental pressures often do not receive compensation, as they are not classified as refugees since environmental causes are not typically considered within the definition of refugee status. Recognizing this gap, we undertook an extensive empirical study at the grassroots level, focusing on individual households. Several villages within the Pathar Pratima block were carefully chosen for data collection through a systematic process outlined in three logical steps. Before delving into these steps, discussing the theoretical underpinnings of environmental migration is essential, as elucidated by Black et al. (2011). Black et al. (2011) identifies five key drivers—environmental, political, demographic, economic, and social—influencing migration decisions at various levels. He emphasizes that environmental factors can disrupt existing political, demographic, economic, and social conditions, indirectly influencing migration decisions—a dynamic that poses significant research challenges. Consequently, our study is structured into three main components: mapping cyclone-induced hazards, assessing vulnerability in agricultural livelihoods, and analyzing the broader impacts on existing socio-economic phenomena, culminating in identifying migration risk areas. Subsequently, villages were selected based on this framework. Several steps have been considered when selecting the sample villages from the study area, addressed in a flow map (Fig. 4.3).

First Step: Mapping Multi-hazards Risk

Six distinct hazards based on research by Mondal et al. (2021) have been selected To create a comprehensive multi-hazard map. These hazards include cyclonic events, tidal surges, erosion and accretion, embankment breaches, soil salinity, and the extent of inundation caused by cyclone-induced flooding. Utilizing data from the IMD atlas spanning the last 30 years, we digitally mapped cyclonic tracks using the Arc GIS platform. Each year's track was buffered by 25 km, and the frequency and intensity of tracks crossing through the Mouza were tallied over the three decades to produce the cyclonic hazard map. Tidal surge height data from 2018 to 2020 were collected from hydrological sites and averaged to generate a tidal surge hazard map using the IDW technique in Arc GIS. The embankment map was sourced from the district development report of South 24 Parganas, rectified, and geo-referenced before generating a density map of embankment areas. Erosion and accretion dynamics were calculated at the village level using spatial scenarios of land and water changes

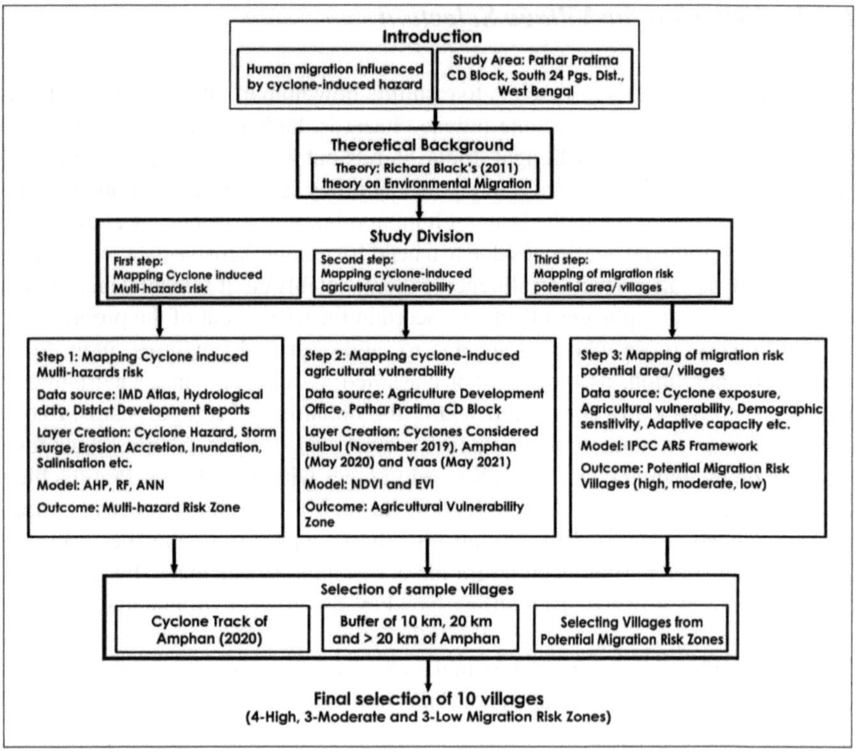

Fig. 4.3 Selecting a village for a sample survey in the flow diagram

over 30 years. Soil salinity data were averaged over the same period. Lastly, the inundation intensity map was created using data from the three most recent cyclones, Bulbul (2019), Amphan (2020), and Yaas (2021), obtained from the Agricultural Development Office in the Pathar Pratima block. Three multi-hazard risk maps were developed using different techniques (AHP, RF, and ANN). After validation, the Random Forest (RF) model was deemed the most accurate and selected for further analysis.

Second Step: Mapping Cyclone-Induced Agricultural Vulnerability

This stage involved gathering secondary data from the Agricultural Development Office in the Pathar Pratima block. The data focused on crop damage caused by Cyclones Bulbul (2019), Amphan (2020), and Yaas (2021) at the individual village level. From this, we computed the total capital losses for villages and, subsequently, the per capita losses for farmers. Additionally, we calculated the average damage across the three cyclones and created a map to illustrate agricultural vulnerability. Furthermore, the secondary data highlighted that those areas inundated by Cyclone Aila in 2009 remained uncultivable for specific villages in the study area that year. This underscores the ongoing challenges faced by southern villages, where, despite a consistent increase in Kharif crop production and area, the cropping intensity for Rabi paddy has steadily declined due to inadequate irrigation facilities and high soil salinity. These findings resulted in a final map depicting the agricultural vulnerability contributing to livelihood crises.

Third Step: Migration Risk Potential Mapping

The aim is to understand the cyclonic vulnerability's direct and indirect effects on the likelihood of human migration using a secondary database. Five key components have been identified to achieve this. The cyclonic exposure, followed by an analysis of the agricultural vulnerability of livelihoods due to cyclones, has been assessed. Then, demographic sensitivity, existing adaptive capacity, and infrastructure were considered. Each component included selected variables, elaborated on in the earlier chapter (Chap. 5). Using the IPSS AR5 methodological framework, the potential impact of cyclones on villages and socio-economic vulnerability has been calculated. Subsequently, they generated a migration risk potential map by combining the five components with assigned weights. Socio-economic factors received the highest weight, reflecting their influence on cyclone-induced migration potential. Agricultural vulnerability and infrastructural vulnerability followed with their respective weightage.

4.3.3 Selection of Sample Villages

A map was created to assess migration risk levels (Fig. 4.4), dividing villages into three classes. The Amphan cyclone track was taken from the IMD atlas map and digitized. During the cyclone, gusty wind speed and intensity were measured at different points on the Pathar Pratima block. Three zones were identified from the Amphan cyclone track, and three buffers were generated for further examination of migration risk areas. These buffers allowed for analysis of how migration risk areas intersected with the cyclone's trajectory. A systematic stratified random selection process was used to select three villages based on their geographical distribution within the buffers and their assigned migration risk levels in the nearest zone of Amphan cyclone. Four villages were selected from a very high migration potential zone, three from a moderate zone, and three from a low migration potential zone. A total of 10 villages were selected for a primary survey (Table 4.1).

Table 4.1 Size of sample from selected villages in Pathar Pratima CD block

Cyclone buffer	Name of the village	Migration risk factor	Total H HHs	Share of HHs to total HHs of the Sample Villages	No sample HHs	Non-migrants HHs ($n = 262$)	Migrants HHs ($n = 488$)			
							1 member	2 members	3 members	>3 members
Within 10 km	Paschim Surendra Nagar	H	1383	12.01	90	23	41	19	5	2
						25.6	*45.6*	*21.1*	*5.6*	*2.2*
	Radha Krishna Nagar	M	656	5.7	43	12	22	7	1	1
						27.9	*51.2*	*16.3*	*2.3*	*2.3*
	Dakshin Gangadharpur	L	2593	22.53	169	37	97	28	5	1
						21.9	*57.4*	*16.6*	*3*	*0.6*
10–20 km	Gobindapur Abad	H	1221	10.61	80	42	27	8	3	0
						52.5	*33.8*	*10*	*3.8*	*0*
	Bishnupur	M	540	4.69	35	17	12	6	0	0
						48.6	*34.3*	*17.1*	*0*	*0*
	Digambarpur	L	1395	12.12	91	22	38	22	8	1
						24.2	*41.8*	*24.2*	*8.8*	*1.1*
Above 20 km	Gobardhanpur	H	243	2.11	16	11	7	3	1	0
						68.8	*43.8*	*18.8*	*6.3*	*0*
	Sridhar Nagar	M	1493	12.97	97	33	45	11	8	0
						34	*46.4*	*11.3*	*8.2*	*0*
	Purba Sripatinagar	H	888	7.71	58	29	18	10	1	0
						50	*31*	*17.2*	*1.7*	*0*
	Purba Surendranagar	L	1099	9.55	71	36	24	8	3	0
						50.7	*33.8*	*11.3*	*4.2*	*0*

(continued)

Table 4.1 (continued)

Cyclone buffer	Name of the village	Migration risk factor	Total H HHs	Share of HHs to total HHs of the Sample Villages	No sample HHs	Non-migrants HHs ($n = 262$)	Migrants HHs ($n = 488$)			
							1 member	2 members	3 members	>3 members
Total		11,511		100	750	262	331	122	35	5
						34.9	*44.1*	*16.3*	*4.7*	*0.7*

Source Computed by the researcher

NB. Data in *Italics* shows the % share; H denotes High, M is Moderate, and L is Low

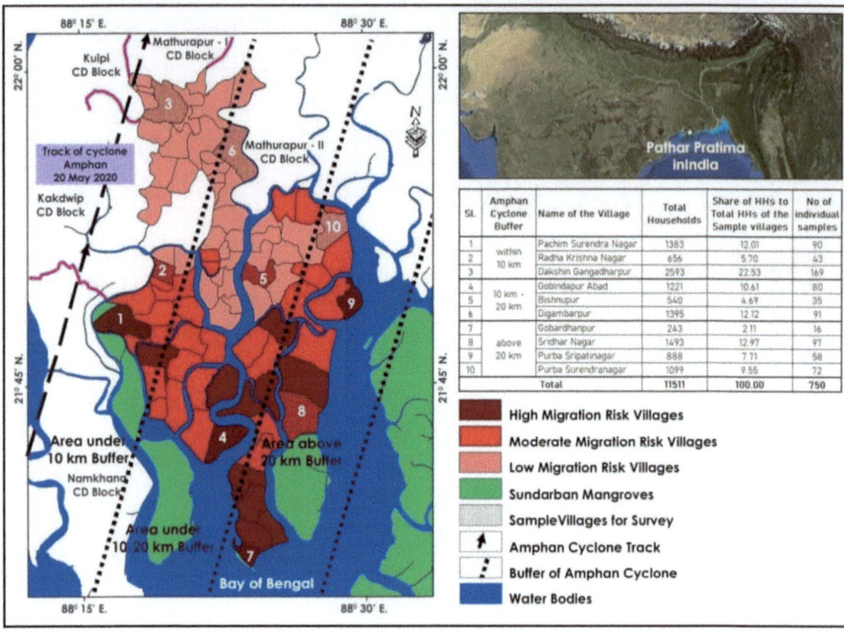

Sl.	Amphan Cyclone Buffer	Name of the Village	Total Households	Share of HHs to Total HHs of the Sample villages	No of individual samples
1	within 10 km	Pachim Surendra Nagar	1383	12.01	90
2		Radha Krishna Nagar	656	5.70	43
3		Dakshin Gangadharpur	2593	22.53	169
4	10 km - 20 km	Gobindapur Abad	1221	10.61	80
5		Bishnupur	540	4.69	35
6		Digambarpur	1395	12.12	91
7	above 20 km	Gobardhanpur	243	2.11	16
8		Sridhar Nagar	1493	12.97	97
9		Purba Sripatinagar	888	7.71	58
10		Purba Surendranagar	1099	9.55	72
Total			11511	100.00	750

Fig. 4.4 Location of sample villages in Pathar Pratima CD Block, India

4.3.4 Sampling Method and Sample Size

In this study, we focused on two main types of households: those who migrated and those who didn't, seeing them as distinct social units. We also looked specifically at migrant households impacted by the Aila and Amphan cyclones but lacked adequate coping mechanisms. This included households where at least one member had temporarily or permanently moved elsewhere to seek additional sources of income, as well as those where no migration occurred. We gathered data from a total of 11,511 households, comprising 4,632 within a 10 km radius, 3,156 within a 10–20 km radius, and 3,723 located beyond a 20 km radius from the area affected by the Amphan cyclone. These households were selected from ten villages severely affected by Amphan.

The sample size was determined using Kothari's formula (2004):

$$n = \frac{z^2 \cdot p \cdot q \cdot N}{e^2(N-1) + z^2 \cdot p \cdot q}$$

where $N = 4632 + 3156 + 3723 = 11{,}511$; $e =$ marginal error (0.01); z value $=$ 2.58; $p = 0.01$ and $q = 0.99$. Employing Kothari's formula, an initial determination of 624 households has been estimated. To ensure a successful study, researchers must provide additional allowances for potential non-response subjects after identifying the minimum required sample size. This means recruiting subjects beyond the minimum required sample size by 20–30% to avoid underestimating the sample size (Bujang and Nurakmal 2016; Bujang 2021). For instance, if a high non-response rate is expected in a self-administered survey, an allowance of 40–50% should be provided. Non-response can occur in various scenarios, such as dropping out or loss to follow-up in cohort studies and experimental studies, as well as missing data or loss of records in observational studies. For example, adding 20% of non-response rate for 624 would require 750 Households for the study. The calculation should be done as follows:

$$n = 624 * 120\% = 750 \text{ Households}$$

Research on migration issues requires a higher confidence level (e.g., 90%, 95%, or 99%) to ensure accurate and reliable results (Bernzen et al. 2019; Carrico and Donato 2019). This is due to the complexity of migration issues, the implications for policy, and the use of comparative analysis. Accurate estimates and trustworthy deductions are crucial for making informed decisions on immigration. Comparative analysis, which benefits from higher confidence levels, is often used to study migrant and non-migrant scenarios. Following that, these 750 households were surveyed. It was found that around 72% of the households in the study are impacted by Cyclone Amphan. Of the households surveyed, 262 (34.93%) were non-migrant, and 488 (65.07%) were migrant households. To account for potential non-responsiveness in sampling, an additional 10% was included, bringing the final sample size to 750 households. A survey was then conducted on these 750 households, representing approximately 72% of the total households affected by cyclone Amphan. Among these surveyed households, 488 (65.1%) were migrant households, while 262 (34.9%) were non-migrant (Fig. 4.5).

4.4 Results

4.4.1 Pattern of Migration

A substantial majority of households (63.10%) opt for intrastate relocation, indicating a preference for shorter, less disruptive moves. In contrast, interstate migration comprises 33.98%, and only a negligible 2.91% of households migrate internationally (Table 4.2). The timing of these migrations is heavily influenced by economic and cultural factors. Festival periods, accounting for 81.55% (Table 4.2) of return migrations, underscore the strong influence of social networks and cultural obligations on migratory decisions. Additionally, significant migration occurs during key agricultural activities such as crop transplanting (10.67%) and harvesting (7.76%), likely driven by the demand for seasonal agricultural labor or financial necessity.

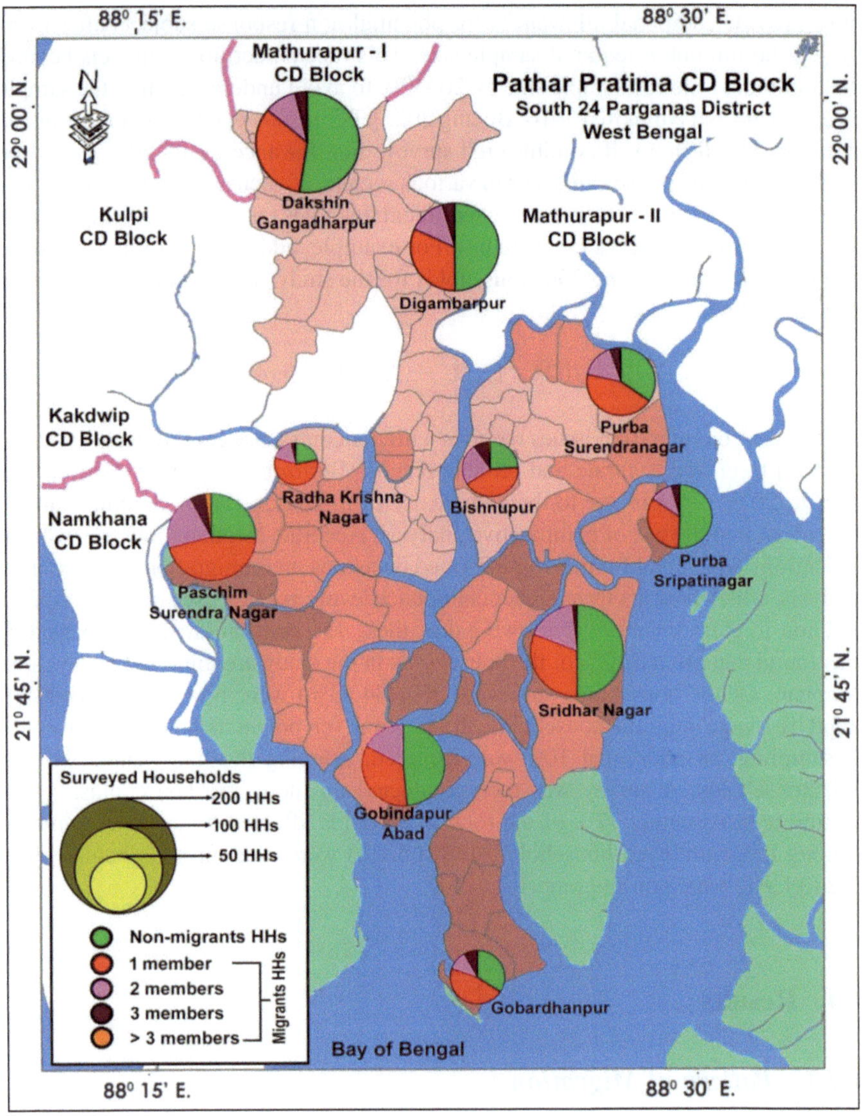

Fig. 4.5 Nature and volume of non-migrants and migrants (no. of member HH^{-1}) in the sample villages of Pathar Pratima CD Block

Return migration varies in frequency, with a notable portion of households making multiple trips back to their hometowns. Evidence suggests seasonal movement or recurrent displacement, with 24.27 and 43.68% of the sample returning to migrate three times and twice, respectively. Furthermore, 32.03% of households return home once a year, potentially as a one-time response to extreme cyclonic events (Table 4.2).

Table 4.2 Descript status of pattern of migration

Variables	Percentage (%)	Variables	Percentage (%)
Age of migrant workers		How many times you are come here in a year?	
15–25 years	28.35	One times	32
26–36 years	38.58	Two times	44
37–47 years	22.83	Three times	24
48–60 years	8.66	Time to return home	
>60 year	1.57	Crop transplanting time	11
Marital status		Crop harvesting time	8
Married	72.97	Festival times	82
Unmarried	27.03	Remittance amount	
Education status of migrant worker		<5000	16
0–5 years	33.94	5000–7000	44
6–10 years	43.12	7000–10,000	24
11–15 years	20.18	>10,000	17
>15 years	2.75	Remittance time	
Education of family members		Monthly	25
0–5 years	45.99	Every couple of months	21
6–10 years	37.12	Every 3 months	29
11–15 years	14.05	Only on special occasions or	24
>15 years	2.84	The decision takes to spend the remittance	
Destination of the migration		By migrant	26
Other country	3	Wife	15
Other state	63	Father	48
Within state	34	Mother	12

Source Computed by the researcher based on primary survey

Migration significantly alters employment prospects and income distributions. Before migrating, the majority of households (63.10%) earned between INR 3,000 and INR 6,000, with a small percentage (0.97%) earning over INR 9,000. After migration, the income distribution shifts, with 3.88% of households earning more than INR 9,000 and 19.41% earning more than INR 6,000 (Table 4.3). This indicates that many households experience an increase in income post-migration, possibly due to better job opportunities in urban or peri-urban areas. Migrants find employment in various industries, including manufacturing (32.03%), construction (36.89%), and other sectors such as stone cutting, textiles, and (Table 4.4) transport (25.24%). The phenomenon of certain migrant households experiencing increased income levels can be attributed to the fact that industries outside of traditional agriculture often offer higher wages.

Migration primarily stems from economic hardships, with 67.96% of households identifying a lack of income as a significant driving factor. Additional factors include

Table 4.3 Share of migrants family in respect to monthly income before and after migration

Monthly income	Before migration	After migration	Monthly income	Before migration	After migration
<3000	24.27	12.62	6000–9000	11.65	19.42
3000–6000	63.11	64.08	>9000	0.97	3.88

Source Computed by the researcher based on primary survey

Table 4.4 Descriptive status of migration types and reasons of migration

Types of migration	Share
Permanent	36
Temporary	64
Working structure of migrant workers	
Construction	37
Decoration	3
Manufacturing	32
Others (stone cutting, textile industry, or transport)	25
Decision of migration	
By family	10
By own	90
Family income change (response by head of the family)	
Yes	62
No	38
Reason for migration	
No source of income	68
Unemployment	22
Asset loss due to a cyclone	6
Failed to reply to the loan	9
Loss due to cyclone and flood	15

Source Computed by the researcher based on primary survey

unemployment (22.33%), loss of assets due to cyclone (5.82%), loan repayment challenges (8.73%) (Table 4.4), and damages from floods and cyclones (14.56%). These factors underscore the economic vulnerability of households in regions prone to cyclonic activity, prompting migration as a survival strategy. Demographically, 38.58% of migrants are aged between 26 and 36, and 28.34% are aged between 15 and 25, indicating that the majority are younger individuals. Moreover, 72.97% of migrants are married (Table 4.2), suggesting that entire families often migrate in search of better prospects or to escape adverse conditions. This age distribution further implies that migration is predominantly a strategy adopted by working-age individuals and their families. Upon migrating, households commonly engage in activities such as house repairs (46.36%) and off-farm employment (2.72%) to rebuild and stabilize their lives. However, migrant households face significant challenges, including low income (65.67%) and asset loss (16.41%), which hinder their efforts to improve their living standards. In contrast, non-migrant households also suffer from low incomes (77.77%) but report lower incidences of asset loss, highlighting that while migration can offer opportunities for higher earnings, it also involves considerable risks and uncertainties. It is noteworthy that most family members who migrate do so temporarily, returning home annually during crop transplanting and harvesting seasons. Only about 30% of households migrate permanently, while the remaining 70%, comprising 225 households, migrate on a temporary basis (Table 4.4).

4.4.2 Demographic and Socio-economic Status of Coastal Dwellers

4.4.2.1 Demographic Composition of Household Participants

A detailed examination of the demographic profile of migrants reveals that the majority are working people, with 42.89% aged (Fig. 4.6) between 16 and 60 years, and 32.51% aged less than 15 years. This data highlights a significant trend: young adults (Fig. 4.7) and individuals in their prime working years are the predominant participants in migration. Additionally, a substantial proportion of migrants, 72.97%, are married. This suggests that migration is often a family-oriented process.

Furthermore, the age distribution suggests that working-age people and their families are the ones that use migration as a strategy the most frequently. The age distribution further implies that migration is most frequently adopted migration as a strategy by working-age individuals and their families. The size of a family plays a crucial role in the decision to migrate. Larger families can afford to have one or more members work outside the home, while smaller families may face more significant challenges due to social security concerns. The presence of multiple family members at home can alleviate some of these concerns, facilitating the decision to migrate. In our survey, the number of family members ranged from as few as one to as many as nine.

The household was categorized based on size to better understand the factors influencing migration. Households with three or fewer individuals are classified as low-size households, those with four to six members as medium-sized households, and those with more than six members as large families. According to the survey, 57% of all sample households fall into the medium-sized category, having four to six members. In contrast, 30% of households are classified as low-size, and 13% are considered large families. This household size distribution provides insight into migration dynamics, indicating that medium-sized families are the most prevalent among migrants, likely due to a balance between having enough members to support the migration process and not being so large as to face insurmountable logistical challenges. The socio-economic vulnerabilities of Scheduled Castes (SC) and Scheduled Tribes (ST), known to be the weaker segments of society, can exacerbate the impacts

Fig. 4.6 According to one of the migrants, the pace of out-migration from the community increased noticeably after Cyclone Amphan. He noted that the majority of the migrants were young individuals who were forced to flee in pursuit of stability and work because the destruction caused by the cyclone had seriously hampered local livelihood possibilities. *Source* By authors on 18-11-2023

Fig. 4.7 Three young adults who work in Kerala as migrant workers in the construction sector. *Source* By authors on 10-01-2024

of any hazards they face. In the study region, the demographic composition of households is as follows: 48% belong to the General category, 34% to the SC category, 4% to the OBC-B (Other et al.) category, and 14% to the OBC-A category. Regarding religious affiliation, 92% of the households are Hindu, while the remaining 8% are Muslim or belong to other minority groups.

The study reveals that approximately 55% of the households are joint families, while 50% are nuclear families. A notable disparity in literacy rates is observed between migrant and non-migrant households. Illiteracy is significantly higher among migrant households, with about 34% being illiterate compared to 20% in non-migrant households. Regarding educational attainment, around 27% of members from migrant households have completed primary school (up to the fifth grade), compared to only 7% from non-migrant households. Additionally, a higher percentage of respondents from migrant households (5.63%) have completed secondary education (up to the tenth grade) compared to those from non-migrant households (4.35%). Historical research indicates that migration rates are higher among joint families than nuclear families. Specifically, 55.1% of households are nuclear, and 44.9% are joint families. Table 4.5 provides a comprehensive summary of the demographic profile of respondents within these households, highlighting the intricate socio-economic factors influencing migration patterns and the educational disparities between migrant and non-migrant households.

4.4.2.2 Economic Status of Coastal Dwellers

Multiple income-earning members within households can significantly bolster our initiative, indicating that more individuals have relocated and successfully established livelihoods to support their families. The study reveals that approximately 43% of sampled households have at least one wage earner. Among the remaining households, 35% have two wage earners, and 22% have three or more financially successful members, as detailed in Table 4.6. This data suggests that migration has facilitated the economic stability of these families. Further studies underscore that migration not only contributes to economic stability but also enhances the capacity of these families to adapt to adverse environmental conditions, such as cyclones and other coastal hazards. To understand this adaptive capacity, researchers employed two key indices: the structure of houses before relocation and after migration. Before migration, most of the population lived in huts or mud houses. However, post-migration, the percentage of families living in such rudimentary structures decreased to 43%. Initially, 26% of households were fully cemented, and 19% were semi-cemented. Following migration, these figures increased to 36% for semi-cemented houses and 30% for fully cemented structures.

Table 4.5 Demographic profile of the respondent households

Demographic determinants	Migrant HHs		Non-migrant HHs		Total		Chi-square (χ^2)
	n	%	*n*	%	*n*	%	
Household type	488	65.07	262	34.93	750	100	
Age group							
0–15 years (young dependent)	881	23.51	337	8.99	1218	32.51	*69.26***
16–60 years (working group)	1095	29.22	512	13.66	1607	42.89	
above 60 years (old dependent)	513	13.69	409	10.92	922	24.61	
Social categories							
Unreserved	210	28	150	20	360	48	*75.67**
SC	217	28.93	38	5.07	255	34	
OBC (A and B)	61	8.13	74	9.87	135	18	
Religion							
Hindu	477	63.6	213	28.4	690	92	*40*
Islam	11	1.47	49	6.53	60	8	
Household size							
1–3 persons	52	6.93	21	2.8	73	9.73	*78.29***
4–6 persons	377	50.27	136	18.13	513	68.4	
7–9 persons	59	7.87	105	14	164	21.87	
Education attainment							
Illiterate	1254	33.47	733	19.56	1987	53.03	52.32**
Up to the primary level	997	26.61	259	6.91	1256	33.52	
Secondary level	211	5.63	163	4.35	374	9.98	
Above secondary level	27	0.72	103	2.75	130	3.47	
Family types							
Joint	312	41.60	101	13.47	413	55.1	4.96**
Nuclear	176	23.47	161	21.47	337	44.9	

Note ** $p < 0.01$; *** $p < 0.001$, *Source* Computed by the researcher based on primary survey

4.4.2.3 Physical Infrastructure

Traditional homes in coastal areas, called Kaccha houses, were constructed with non-brick materials such as bamboo, wood, plant material for roofs, and clay for walls. In contrast, some homes were pucca, built with more durable materials like bricks and wood from the Sundarbans mangrove forest, while others were semi-pucca, featuring tin roofs and walls. These distinctions are critical in assessing the resilience of housing structures. Table 4.7 provides a comparative analysis of the housing conditions of

Table 4.6 *Economic* status of the households

	Migrated	%	Non-Migrated	%	Total	%	Chi-square
One member earner	283	37.73	124	16.53	407	54.27	10.1**
Two-member earner	120	16.00	81	10.80	201	26.80	
Three-member earner	85	11.33	57	7.60	142	18.93	

Note ** $p < 0.01$; *** $p < 0.001$
Source Computed by the researcher based on primary survey

Table 4.7 Housing structure of migrant and non-migrant households before and after Amphan

	Before Amphan				After Amphan			
	Migrant HHs		Non-migrant HHs		Migrant HHs		Non-migrant HHs	
	n	%	*n*	%	*n*	%	*n*	%
Kaccha	327	67.01	102	38.93	117	23.98	59	22.52
Semi-pucca	106	21.72	63	24.05	238	48.77	78	29.77
Pucca	55	11.27	97	37.02	133	27.25	125	47.71
χ^2	79.688**				35.496**			

Note ** $p < 0.01$; *** $p < 0.001$
Source Computed by the researcher based on primary survey

households with and without migrants before and after Cyclone Amphan. The data indicate a significant shift in housing types post-cyclone.

There was a marked decline in the number of Kaccha houses and a corresponding increase in the number of semi-pucca and pucca constructions. This trend was observed among migrant and non-migrant households, underscoring the broader impact of migration on improving housing resilience. The transition from Kaccha to more durable housing structures such as semi-pucca and pucca reflects an adaptive response to environmental challenges. Migrant families, in particular, have leveraged their improved economic status to invest in better housing, enhancing their ability to withstand future cyclonic events.

4.4.2.4 Occupational Structure of Coastal Dwellers

This shift indicates improved living conditions and demonstrates the critical role of economic mobility and migration in fostering community resilience in coastal areas. This section examines the financial repercussions on households following Cyclone Amphan in 2020, focusing on the differences between migrant and non-migrant households in terms of occupation, income, and debt status. Agriculture has traditionally been the primary source of income for the majority of these households, with approximately 75% engaged in farming, forestry, animal husbandry, and fishing (Fig. 4.8), as reported by the Census of India in 2011. Many small-scale farmers, particularly those residing along the coast, depended heavily on rain-fed agriculture, primarily cultivating Aman paddy. Additionally, shrimp aquaculture was a dominant practice in fisheries, alongside significant reliance on other fish and forest resources. Cyclones and other climate-related events severely impact agricultural productivity, with smallholders and marginal farmers bearing the brunt of these effects. Field survey data, as presented in Table 4.8, illustrate the changes in occupation for both migrant and non-migrant households before and after Cyclone Amphan. Before the cyclone, about 56% of migrant and 27% of non-migrant household members were employed in agriculture and in the post-cyclone, these figures dropped to 28% and 20%, respectively.

Conversely, the proportion of non-farm wage workers increased significantly, rising to 60% among non-migrant households and 54% among migrant households. Notably, involvement in fishing declined to approximately 2.0–2.5% for both categories of households. Climate change threats, such as storm surges, floods, cyclones, droughts, and temperature variations, contribute to water and soil Salinization, diminishing agricultural yields in coastal areas. Cyclone Amphan exacerbated this issue

Fig. 4.8 "How I depict my fear by the way my heart races in terror. My thoughts are constantly focused on when I can return to open spaces," shares Sabita Das, a 41-year-old crab hunter and fisherwoman. She describes the chilling fear she experiences while searching for crabs in the dense mangrove forests of the Sundarbans. Throughout the crab-hunting season, she maneuvers a boat along the rivulets and creeks, remaining in a state of heightened alert, always mindful of the potential presence of lurking tigers. *Source* By authors on 10-01-2024

Table 4.8 Change in occupation by migrant and non-migrant households before and after Amphan

	Migrant HHs				Non-migrant HHs			
	Before Amphan		After Amphan		Before Amphan		After Amphan	
	n	%	*n*	%	*n*	%	*n*	%
Farm-based activity	273	55.94	136	27.87	71	27.10	53	20.23
Non-farm activity	113	23.16	264	54.10	126	48.09	157	59.92
Fishing activity	87	17.83	74	15.16	23	8.78	17	6.49
Service	5	1.02	5	1.02	12	4.58	13	4.96
Business	10	2.05	9	1.84	30	11.45	22	8.40
Average monthly income (Rs. HH^{-1})	8478.00		6256.00		10,587.00		9879.00	

Note ** $p < 0.01$; *** $p < 0.001$
Source Computed by the researcher based on primary survey

by breaching an embankment, allowing saline water to inundate farms and shrimp aquaculture operations. This led to prolonged water logging and increased salinity levels, rendering agricultural land less productive and reducing crop yields. The decline in agricultural productivity and the shift toward non-farm wage labor underscore the profound impact of Cyclone Amphan on coastal households. The increased salinity and waterlogging have compromised the immediate agricultural output and posed long-term challenges to the viability of farming in these regions. This situation necessitates reconsidering livelihood strategies, emphasizing the need for diversified income sources and enhanced resilience against climate-induced adversities.

4.4.3 Cyclonic Exposure and Risk

The region's vulnerability to cyclones and storm surges and its proximity to rivers or the ocean pose significant risks to residents' assets, which can be damaged or lost during such events (Fig. 4.9). To analyze the factors influencing migration decisions, we have categorized household distances from the river or shoreline into three groups: less than 500 m, between 500 m and one kilometer, and more than one kilometer. Among the sampled families, 476 households, representing 63.47% of the sample, are within 500 m of the coast or riverine area. Conversely, only 15.60% of households fall into the third group, situated more than one kilometer from the shoreline (Table 4.9).

The distribution of households by physical location is 44% in coastal villages, 23.2% in riverine villages, and 33% in inland villages. These distributions highlight the varied exposure levels of households to coastal and riverine hazards. Numerous

Fig. 4.9 Despite severe storms like Amphan, Bulbul, and Yaas, coastal residents tend to stay near the shore because of an emotional sense of place attachment. *Source* By authors on 21-06-2020

studies have demonstrated that migration is a critical adaptation strategy for people living near the ocean. The duration of hazardous environmental conditions following a cyclonic storm underscores the area's vulnerability. In the aftermath of the cyclone, we measured the "duration of water logging" to assess the persistence of these adverse conditions. Approximately 32.13% of household heads reported that water logging lasted for nearly one month, while 46% indicated that it persisted for two weeks (as detailed in Table 4.9). The proximity of households to the coast or river significantly impacts their decision to migrate, as those closer to these bodies of water are more likely to experience severe and prolonged water logging, prompting them to seek safer locations. The data reveals a clear pattern: Households within 500 m of the shoreline or rivers are disproportionately represented among those choosing to migrate, reflecting their heightened vulnerability to cyclonic impacts and water logging. In summary, the analysis highlights the critical role of geographic proximity in influencing migration decisions. Households closer to the coast or river are more exposed to cyclonic and storm surge risks, leading to more significant asset damage and prolonged water logging. As a result, migration emerges as a viable adaptation strategy for these households, enabling them to mitigate the risks associated with living in high-risk areas. This underscores the importance of considering geographic factors and environmental conditions in developing effective policies and interventions to support vulnerable communities in coastal and riverine regions.

Table 4.9 Exposure and risk to cyclone in Coastal Sundarbans

	Migrated household	%	Non-migrated household	%		Total %	Chi-square
Distance to river/coastline							
Less than 500 m	388	51.73	88	11.73	476	63.47	78.5***
500 m–1 km	35	4.67	122	16.27	157	20.93	
More than 1 km	65	8.67	52	6.93	117	15.60	
Location of household							
Riverine	138	18.40	36	4.80	174	23.20	7.14**
Coastal	313	41.73	15	2.00	328	43.73	
Inland	37	4.93	211	28.13	248	33.07	
Inundation surrounding times							
One week after the cyclone	67	8.93	97	12.93	164	21.87	33.4***
Two weeks after the cyclone	224	29.87	121	16.13	345	46.00	
More than 3 weeks after the cyclone	197	26.27	44	5.87	241	32.13	

Note ** $p < 0.01$; *** $p < 0.001$
Source Computed by the researcher based on primary survey

4.4.4 Livelihood Crisis

Cyclonic threats substantially indirectly influence migration in coastal regions, primarily due to the interplay between these natural disasters and prevailing socio-economic conditions (Fig. 4.10). A critical factor driving this migration is cyclones' severe impact on the livelihoods of coastal inhabitants, who often rely on agriculture and fishing for their sustenance. To quantify these impacts, two indices were considered: crop losses and fishing output reductions attributable to cyclonic activity. The extent of crop loss varies significantly among households, depending on the land they own. Data reveals that families with smaller holdings, producing between one and five quintals of crops, experienced devastating losses, with approximately 94.27% of their produce destroyed. In contrast, households with more extensive holdings, producing five to ten quintals or more, reported substantially lower losses, at about 4.93% and 0.80%, respectively, for migrant and non-migrant households. This disparity underscores the heightened vulnerability of smaller landholders to cyclonic damages.

On the other hand, the impact on fishing outputs presents a different picture. Around 30.80% of households indicated that their fishing productivity remained unaffected, primarily due to insufficient land for fishing cultivation. However, the

Fig. 4.10 Majority of the people depend on natural resources in Sundarbans. *Source* By authors on 10-01-2024

majority still suffered significant financial losses due to storm surges. Specifically, over 50.13% of households incurred losses of up to 10,000 rupees in fishing production, while 19.07% faced even steeper losses exceeding 10,000 rupees. The recurrent devastation caused by cyclone has compelled many residents to migrate or adapt to their living conditions. According to Focus Group Discussions (FGD), many individuals have relocated to more secure, concrete houses, known locally as Pucca homes, to mitigate the persistent damage to their traditional mud dwellings. This shift highlights a community's resilience and adaptive strategies in the face of continual cyclonic threats. Survey data reveals that the extent of housing damage approximately 38.53% were partially destroyed, 39.47% were utterly devastated, and only a minority of fully cemented homes remained largely unharmed. Table 4.10 illustrates the critical importance of durable housing in enhancing resilience against cyclonic hazards, influencing migration patterns as they live in low, less durable houses.

Table 4.10 Livelihood crisis due to cyclonic damage

	Migrant HHs		Non-migrant HHs		Total		Chi-square
	n	%	*n*	%	*n*	%	
Dwelling house damaged							
Fully	*189*	25.20	107	14.27	296	39.47	14.9***
Partially	154	20.53	135	18.00	289	38.53	
No damaged	145	19.33	20	2.67	165	22.00	
Crop damage due to storm surge inundation							
One to five quintals	362	48.27	210	28.00	572	76.27	137***
Five to ten quintals	89	11.87	46	6.13	135	18.00	
More than ten quintals	37	4.93	6	0.80	43	5.73	
Damaged Fishing production due to cyclone							
No damaged	127	16.93	104	13.87	231	30.80	83.3***
Up to 10,000 rupees	245	32.67	131	17.47	376	50.13	
10,000–20,000 rupees	69	9.20	12	1.60	81	10.80	
More than 20,000 rupees	47	6.27	15	2.00	62	8.27	
Average loss at the Household level	3987.5		2879				

Note ** $p < 0.01$; *** $p < 0.001$
Source Computed by the researcher based on primary survey

4.4.5 Natural Capital

Land ownership shapes rural migration patterns, reflecting deeply ingrained socio-economic structures. This relationship is particularly pronounced among disadvantaged social groups, such as Scheduled Castes (SC) and Scheduled Tribes (ST), who face severe limitations in land ownership compared to other segments of society. Landholders are categorized into three groups based on the extent of their landholdings: landless individuals, who possess no land for cultivation; marginal landholders, who own less than two Bigha; and small landholders, who hold more than two Bigha. A survey encompassing 750 households revealed that 44% of participants are landless, meaning they do not have access to land for farming and frequently have to rent land to meet their needs. Among those classified as small landholders who own more than two Bigha, 43% of the households are categorized as marginal landholders. This data underscores a critical distinction between landholding capacities and the implications for rural livelihoods (Table 4.11).

The survey further dissects landownership across various social categories, revealing stark disparities. Less than 5% of families from the general category are landless (Fig. 4.11), highlighting their relative stability regarding land access. Conversely, the majority of SC families, approximately 20.4%, are landless, reflecting a significant socio-economic disparity. The distribution of small landholders with more than two Bigha farmland also reveals imbalances: only 8.85% of this category are from SC backgrounds, while the representation from Other Backward Classes (OBC) is also notably low, with 1.4% from OBC-B and 2% from OBC-A. The chi-square test applied to these findings indicates a statistically significant difference between landholding categories and household types, underscoring the entrenched inequalities in land distribution. This significant discrepancy highlights how access to land—or the lack thereof—can influence migration decisions and perpetuate social and economic inequalities within rural communities.

Table 4.11 Landholding capacity according to social category

Social categories	Landless		Marginal landholder		Small landholder		Total	
UR	32	4.3%	169	22.5%	66	8.8%	282	35.6%
SC	150	20.0%	91	12.2%	8	1.0%	255	33.2%
OBC-B	61	8.2%	5	0.7%	15	2.0%	82	10.9%
OBC-A	107	14.3%	41	5.4%	5	0.7%	153	20.4%
Total		43.8%		43		13		100%
Chi-square (χ^2)	27.6***							

Note ** $p < 0.01$; *** $p < 0.001$
Source Computed by the researcher based on primary survey

Fig. 4.11 Bound by tradition: despite cyclone damage, loyalty to ancestral land prevails. *Photo* By authors on 19-04-2023

4.4.6 *Obtaining a Loan, Its Purpose, and the Challenges Faced by Both Migrant and Non-migrant Households Following the Amphan Cyclone*

Table 4.12 provides a detailed comparison of how migrant and non-migrant households managed their financial challenges following the devastation of Cyclone Amphan in 2020 (Fig. 4.12). According to the data, approximately 87% of migrant households and 68% of non-migrant households sought financial assistance through loans from various sources. The average loan amount for migrant households was INR 10,988, notably higher than the INR 7,545 average for non-migrant households. This disparity indicates that migrant households may have faced more financial strain or required more substantial support to cover their losses.

Despite the significant uptake of loans, only 8.5% of migrant households and 67.8% of non-migrant households managed to repay their debts after the cyclone. Most households struggled with monthly repayments, reflecting broader financial

Table 4.12 Loan obtained, purpose of loan, and problems of repayment by migrant and non-migrant households after Amphan cyclone

	Migrant HHs	Non-migrant HHs	Chi-square (χ^2)
	%	%	
Obtain the loan: Average amount of loan in INR	86.68	67.56	
Loan used >50% for repairing/maintaining (housing and sanitation)	37.59	70.06	63.17**
Loan used >50% for livelihood	27.66	11.86	
Loan used >50% for food	21.04	1.69	
Loan used >50% for education	5.91	7.34	
Loan used >50% for health issues	7.8	9.04	
Problems of loan repayment	91.49	32.2	32.01**
Due to low income	70.03	57.89	
Due to asset loss	23	36.84	
Due to increased household expenditure	6.98	5.26	

Note ** $p < 0.01$; *** $p < 0.001$ (*Source* Computed by the researcher based on primary survey)

difficulties exacerbated by the disaster. Focus group discussions revealed that loans were often used for diverse purposes. Some households invested borrowed funds in renovating their homes or financing non-farm income-generating activities.

These included purchasing vehicles such as cars, rickshaws, vans, boats, fishing nets, power tillers, and cattle. Table 6.9 highlights that approximately 38% of migrant households and 70% of non-migrant households used their loans primarily for maintaining or repairing their housing assets. In contrast, 28% of migrant households and only 12% of non-migrant households directed their loans toward subsistence-related activities, such as agriculture and fish farming. Using loans for different purposes reveals a stark contrast between the two groups. Notably, loans for health-related issues were utilized by a significantly lower proportion of migrant households (about 8%) than non-migrant households (approximately 9%). That means health-related expenses were a relatively minor factor in migrant loan utilization. The post-disaster phase saw loans playing a crucial role in helping affected communities mitigate financial losses and generate income. However, repaying these loans proved challenging for many, revealing underlying financial vulnerabilities. The study underscores that coastal communities faced various barriers to repaying their debts, reflecting broader socio-economic challenges and the financial strain induced by the cyclone.

Fig. 4.12 **a** From the Mangroves to home: Sundarbans Crabbers' return journey, **b** Beyond farming: Amphan cyclone prompts occupational change for rural workers, **c** From early childhood playing with anchor of a boat *Source* By authors on 30-05-2021

4.4.7 Drinking Water, Sanitation, and Hygiene (WASH) Scenario

In the wake of Cyclone Amphan in 2020, the coastal communities of Pathar Pratima faced severe and compounded challenges, particularly concerning water scarcity and salinity intrusion into groundwater sources.

The health crisis in the post-cyclone period was made worse by flooding-related contamination of drinking water sources (Fig. 4.13), inadequate sanitation, and interruptions in healthcare services. These communities, heavily reliant on surface water for their daily sustenance, found themselves grappling with a crisis in water access. Migrant and non-migrant households predominantly drew water from tube wells

(Fig. 4.14) and small, isolated ponds. Survey data highlighted this dependency: 35% of non-migrant households and 44% of migrant households relied on tube wells, while 28% of non-migrant households and 32% of migrant households depended on pond water. The cyclone's intense tidal surges wreaked havoc on these crucial water sources. The inundation of low-lying areas resulted in the saturation of soil and water with high levels of salinity, which rendered both tube wells and pond water unsuitable for consumption and domestic uses such as cooking, bathing, and washing. This contamination left residents in a precarious situation, severely compromising their access to clean water. Before the cyclone, sanitation coverage in the affected areas was relatively robust, at approximately 80%, with households utilizing hygienic toilet facilities. However, the cyclone's destructive force undermined these facilities, many of which were constructed from mud, straw, or plastic. As a result, only the brick-built latrines, which were also eventually compromised by the flooding, remained partially intact. In response to the immediate need for sanitation, non-governmental organizations (NGOs) provided temporary pit latrines. Nevertheless, allocating one latrine for every 15 households proved grossly inadequate, leading some residents to resort to unhygienic alternatives such as river water or makeshift brushwood setups.

Focus group discussions revealed the dire reality faced by these communities. The acute shortage of potable water was compounded by the deteriorating sanitation conditions, leading to a surge in waterborne and skin diseases, including diarrhea, cholera, and food poisoning. These health issues disproportionately affected vulnerable groups, particularly women and children. The increased salinity levels exacerbated the situation, contributing to a significant rise in such ailments. Alarmingly, around 40% of the population succumbed to waterborne diseases, underscoring the

Fig. 4.13 Health-related problems become a key worry for the impacted community during and following cyclonic risks. Specifically, a considerable proportion of individuals reported experiencing waterborne illnesses, with diarrhea being the most common. *Source* By authors on 18-11-2023

Fig. 4.14 Cyclone damaged and contaminates the drinking water, and for that govt. installed a new tubewell. *Source* By authors on 18-11-2023

severity of the crisis. In the broader context of Indian communities, women often remain in climate-vulnerable areas while men migrate in search of employment opportunities. As a result, women bear the primary responsibility for managing daily household logistics and are disproportionately affected by climate-induced adversities. The compounding effects of water scarcity, salinity intrusion, and deteriorating sanitation have placed an immense burden on these women, highlighting their critical role and the severe impact of climate change on their daily lives.

4.4.8 Relief Work After Cyclone Amphan to Migrant and Non-migrant Households

The aftermath of major disasters often leaves communities in profound distress, necessitating a coordinated and robust response from various external entities to help them recover and rebuild. In the wake of Cyclone Amphan's devastating impact, many organizations mobilized to assist the affected populations, reflecting a comprehensive and multifaceted humanitarian response. Immediately following the cyclone, both the Government of West Bengal (GoWB) and the Government of India (GoI) were at the forefront of relief efforts, alongside numerous Non-Governmental Organizations (NGOs) and humanitarian agencies from both national and international spheres. These entities acted swiftly to deliver essential aid to the communities grappling with the aftermath of the disaster. Their support encompassed various relief supplies, including food, clothing, medicines, and other critical resources necessary for survival and recovery. This relief effort was sustained for up to a year following the cyclone, underscoring the prolonged and ongoing nature of the recovery process. The government's response was particularly comprehensive, extending beyond immediate relief to include a range of humanitarian aids such as emergency shelter, potable water, financial assistance, and continued provisions of essential goods. Specifically, under the National Disaster Mitigation programs, approximately 344 households in the selected villages of Pathar Pratima were allocated monthly rations of 5 kg of rice per person. This support continued until November 2020, highlighting the commitment to ensuring food security for an extended period following the disaster.

In Pathar Pratima, the relief efforts were multifaceted, addressing various needs of both migrant and non-migrant households. The provided aid included food assistance, clothing, water and sanitation supplies, emergency medical aid, shelter materials, tents, and cash grants. This comprehensive approach addressed immediate needs while laying the groundwork for longer-term recovery. Additionally, the Chamber Secretariat played a crucial role in distributing essential items. Their efforts included delivering 500 tarpaulin sheets, critical for temporary shelter; 500 solar lanterns, which provided much-needed illumination in the absence of electricity; and essential medicines to address health concerns exacerbated by the disaster. Furthermore, twelve thousand packets of biscuits were distributed, offering a supplementary food source to the affected individuals. These items were handed over to the local Block Development Officer, ensuring their effective distribution within the community. The collective actions of these diverse entities reflect a well-orchestrated humanitarian response designed to address both immediate and ongoing needs in the wake of Cyclone Amphan. The integration of relief efforts, from food and medical aid to shelter and sanitation, illustrates the comprehensive strategy required to support communities recovering from such profound devastation (Fig. 4.15).

Table 4.13 presents a comprehensive overview of the relief items (Fig. 4.16) distributed to migrant and non-migrant households in response to a significant disaster. The table illustrates that aid was uniformly provided across both groups, encompassing essential resources such as food supplies, water and sanitation provisions, shelter materials, and livelihood support. This systematic distribution was influenced mainly by the extent of damage sustained by each household. A closer examination of the data, particularly regarding migrants from the coastal Sundarbans region, reveals critical insights into their unique needs and vulnerabilities. The Sundarbans, a coastal area, is highly susceptible to natural calamities like cyclones and the adverse effects of rising sea levels. Such environmental hazards often compel the residents of this region to migrate in search of safer living conditions and better employment opportunities. Economic displacement, environmental degradation, and the broader impacts of climate change, broader impacts frequently drive the migration from Sundarbans. These factors contribute to an elevated need for fundamental

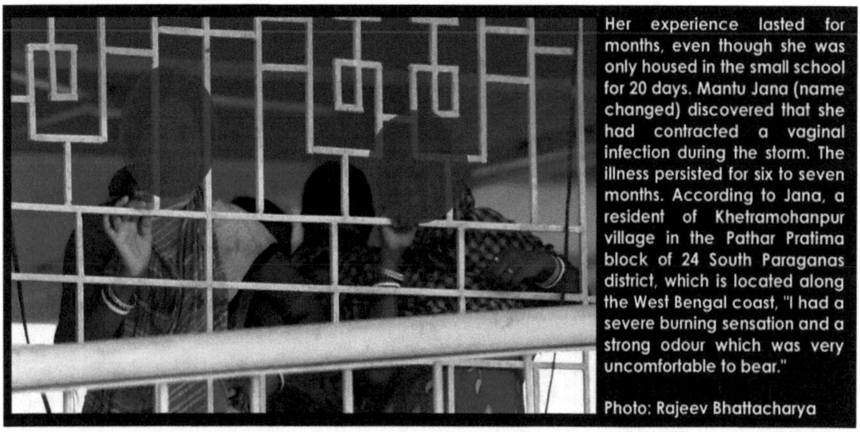

Fig. 4.15 Stranded by cyclone Amphan: women's menstrual health crisis in South 24 Parganas. *Source* By Rajeev Bhattacharya

services, including food, shelter, water, sanitation, medicine, and livelihood support among these migrants. Relocating to a new environment can present significant challenges for these individuals, necessitating a tailored approach to address their specific needs effectively. To adequately meet the needs of migrants from the Sundarbans, policies and interventions must be carefully crafted to consider their distinct vulnerabilities. These measures should enhance their adaptability and well-being in new settings. Such an approach would involve addressing immediate needs and facilitating a smoother transition for migrants as they integrate into new communities.

In addition to the relief items listed in Table 4.13, Focus Group Discussions (FGDs) revealed that each affected household received an additional sum of approximately Rs. 5000 to address their immediate needs. However, this financial aid was insufficient to recover from the losses and damages incurred due to the disaster. As a result, some households were prompted to consider migration as a viable option to improve their situation. Following the distribution of emergency aid, various

Table 4.13 Relief and assistance received by migrant and non-migrant households after Amphan

	Migrant HHs		Non-migrant HHs		Total	
	n	%	*n*	%	*n*	%
Food	288	38.40	56	7.47	344	45.87
Shelter	193	25.73	87	11.60	280	37.33
Water	142	18.93	45	6.00	187	24.93
Sanitation	178	23.73	66	8.80	244	32.53
Medicine	221	29.47	78	10.40	299	39.87
Livelihood	56	7.47	23	3.07	79	10.53

Source Computed by the researcher based on primary survey

Fig. 4.16 An image of a relief camp during Cyclone Amphan in 2020, where hygiene and sanitation made significant compromises. *Source* By authors on 30-05-2020

governmental bodies, NGOs, and donor agencies launched rehabilitation programs to rebuild the livelihoods of the affected communities. Despite these efforts, the short-term relief efforts, which included the establishment of community kitchens by HEAL in severely affected areas and distributing rations in other locations, still needed to fully meet the households' expectations. Households anticipated long-term relief or recovery assistance, such as wage employment opportunities, rather than transient support measures. This expectation could have influenced some affected household members to consider migration to secure a more stable future. While the immediate relief aid alleviated urgent needs, its temporary nature might inadvertently lead to dependency if not complemented by sustained, long-term support. In summary, while Table 6.10 provides a detailed account of the relief distribution, it underscores the need for a nuanced approach to aid, especially for migrant households from vulnerable regions like the Sundarbans. Effective policies should address immediate and long-term needs to foster resilience and support sustainable recovery for affected individuals and communities.

4.5 Discussions

The Sundarbans are highly susceptible to various hazards. Every year, at least one cyclone strikes the coastal areas, causing significant disruptions by damaging the inhabitants' living conditions and livelihoods. Many makeshift houses are either partially or destroyed. Most people are primarily engaged in basic activities for their sustenance. The intrusion of saline water inundates the farmlands, breaching the earthen embankments that are vital for the people of the Sundarbans.

Consequently, some individuals are forced to migrate elsewhere for labor opportunities to support their livelihoods. In response to this challenging situation, we have undertaken an empirical study. Initially, we developed a structured questionnaire focusing on cyclone damage and various socio-economic conditions. This questionnaire was used to gather data from both migrant and non-migrant households. The sample size was determined using a standard equation, resulting in a survey of 750 households.

In broad terms, demographic analyses indicate that younger men tend to be more inclined toward migration. Numerous studies suggest that men, often fulfilling traditional roles as primary breadwinners, are likelier to relocate in pursuit of cash earnings. The migrant goes outside and leaves women, children, and the elderly behind in rural areas, with migrants often sending back remittances to support their families (Falk 2015; Swain 1996; Swapan and Gavin 2011). Environmental challenges faced by rural households can catalyze migration, creating a push factor (Falk 2015; Swain 1996; Swapan and Gavin 2011). Climate change and weather fluctuations have directly or indirectly influenced human populations' migration patterns (McLeman 2013; McLeman et al. 2008; McLeman and Smit 2006). Climate variability affects various aspects of human life, such as health, financial capital, socio-economic activities, and wages, thereby significantly prompting migration (Dallmann and Millock 2017).

Additionally, Scheduled Castes (SC) and Muslims are found to be more predisposed to migration due to their heightened economic and social vulnerability. The study also highlights that these communities have a more significant proportion of landless individuals than higher caste groups. Notably, the illiteracy rate among migrant households is substantially higher, around 34%, in contrast to non-migrant households, which stand at approximately 20%. Interestingly, the study suggests that literate and illiterate individuals exhibit an equal propensity for migration. However, research by Gray and Mueller (2012) suggests that while flooding events result in significant short-term displacement, there is limited evidence to support the notion of long-term migration resulting from such occurrences.

Notably, almost half of the households who migrated—48.27% to be precise—endured agricultural setbacks averaging around five quintals, while around 16.27% of households who did not migrate faced similar losses. So, the absence of employment opportunities in rural areas compels rural men to seek work outside their villages (Debnath and Nayak 2022). Additionally, about 28.67% of migrant households suffered financial losses in fishing production, amounting to roughly INR

10,000, compared to only 3.33% of non-migrant households encountering such losses. Among impoverished and landless households, younger members are more inclined to migrate for daily wage labor to support their families (Hunter et al. 2015). Further concerns include salt leaching, deterioration of neighboring farmlands, and weakening defenses against storms and floods (Johnson 2018). These findings underscore the heavy reliance of migrant households primarily on agriculture, while non-migrant households leaned more toward fishing or aquaculture, particularly shrimp farming. The expenses incurred per migrant household surpassed those of non-migrant households. The research established that a small percentage of migration, approximately 10%, is directly linked to the impact of hazards, though this group remains highly vulnerable and concerned about the nation's welfare (Bernzen et al. 2019). Conversely, participation in non-farm wage labor increased following cyclones, with 52.5% of migrant households and 60% of non-migrant households engaging in such work. Migrant households earned around INR 6256 monthly, while non-migrant households earned approximately INR 9879—a statistically significant difference.

In the aftermath of a cyclone lasting over two weeks, the likelihood of migration tends to increase, as Gray and Mueller (2012) and Chen et al. (2017) noted. Surprisingly, despite their heightened vulnerability to environmental pressures, impoverished individuals often opt against migration due to resource constraints, as Gray and Mueller (2012) and Chen et al. (2017) highlighted. Insights gleaned from focus group discussions (FGDs) underscored how some individuals resorted to borrowing funds to reconstruct their homes or venture into off-farm income-generating pursuits, such as acquiring vehicles, rickshaws, vans, fishing paraphernalia like boats and nets, power tillers, and livestock. A notable disparity emerged in the allocation of financial resources between migrant and non-migrant households, with approximately 12% of non-migrant households and 28% of migrant households directing their investments toward livelihood-centric endeavors. Particularly striking were the differences observed in using loans for fishing and agriculture ventures. The transition from vulnerable, temporary "Kaccha" housing to more resilient semi-pucca and pucca dwellings reflects a discernible trend aimed at bolstering coping mechanisms. Alarmingly, statistical data revealed that nearly 40% of the populace fell prey to waterborne ailments. In many Indian communities, it is often the women who remain in climate-exposed zones while men seek employment opportunities elsewhere. Consequently, women shoulder the primary responsibility of managing day-to-day affairs, profoundly affected by the adverse repercussions of climate variability. Despite their heightened susceptibility to environmental perils, the impoverished are less inclined to migrate owing to the scarcity of resources, as emphasized by Gray and Mueller (2012) and Chen et al. (2017).

Social capital plays a pivotal role in shaping migration patterns, with individuals residing in households where neighbors offered significant assistance during or after the latest cyclone exhibiting a higher propensity to migrate. Notably, there was no discernible difference in the distribution of grants between migrant and non-migrant households. Relief aid distribution primarily hinged upon the extent of damage sustained by each household. However, insights from Focus Group Discussions (FGDs) divulged that apart from the designated relief provisions, each affected household also received approximately Rs. 5000 to address immediate necessities. Following Cyclone Amphan, landless households experienced an increase in both migrant and non-migrant segments. However, there was a diminishing trend in migration among households possessing landholdings. Furthermore, Akhter and Bauer (2014) asserts that households with larger farms exhibit lower migration tendencies. A notable correlation emerged between migrants' age, educational attainment, and their inclination to migrate. Their adeptness, propensity for risk-taking, extensive social connections, and access to information and employment opportunities collectively heightened their propensity to migrate.

4.6 Concluding Remarks

This study provides a comprehensive analysis of the socio-economic impacts of Cyclone Amphan, which struck the southern coastal regions of West Bengal in 2020, focusing on the aftermath experienced by both migrant and non-migrant households within the Pathar Pratima CD Block of the South 24 Parganas district. The research highlights the profound disruption caused by cyclone, illustrating how it prompted many residents to seek alternative living arrangements in response to the extensive damage inflicted upon their communities. The study reveals a stark and distressing reality: Cyclone Amphan severely impacted livelihoods, income sources, housing stability, food security, and the agricultural sector. The cyclone's ferocious winds and subsequent water logging devastated agriculture, fishing, aquaculture, and shrimp farming. Elevated salinity levels in soil and water bodies, resulting from the floodwaters, led to widespread crop failures and diminished agricultural yields. Crop damage forced many households to diversify their income sources as a coping mechanism. Farmers faced monumental losses just before harvest time, forcing many to migrate for relief and better opportunities.

One of the study's key findings is the noticeable shift in landholding patterns within migrant households. The number of small landholders diminished, while the proportion of functionally landless and marginal farmers increased significantly. This shift reflects the broader impact of the cyclone on land ownership and usage, exacerbated by the collapse of protective embankments. These embankments, crucial for safeguarding agricultural lands and shrimp farms from inundation, were severely damaged, leading to heightened soil salinity and further agricultural distress. The prolonged reconstruction efforts to repair these embankments continued, exacerbating soil salinity even 2 years after cyclone, rendering recovery efforts futile for many farmers. In response to the economic strain, affected households increasingly relied on loans from both formal and informal sources. The study found that migrant households exhibited a significantly higher loan repayment rate than their non-migrant counterparts. The requirement for loans indicates financial resilience among migrants that might be attributed to their diversified income sources or more effective management strategies. The impact on daily life was also pronounced, with a notable decline in the average monthly working days among migrant households. This decrease suggests that migration might have become a strategic response to enhance income or manage liabilities arising from loans. Food insecurity emerged as a critical issue after cyclone, with many individuals enduring severe hunger, including having only two meals a day during the crisis. Despite these challenges, migrant households reported higher food consumption levels than non-migrant households in the aftermath of the cyclone. This difference highlights how migration and associated occupational shifts have significantly influenced dietary patterns and food security.

In conclusion, the research underscores that households engaged in migration or livelihood diversification exhibited greater resilience regarding income stability, housing security, food availability, and loan repayment. The positive side of migration suggests that migration or diversification of livelihoods can be a pragmatic strategy for mitigating the effects of climatic disruptions such as Cyclone Amphan. The study advocates for a nuanced understanding of migration as a coping mechanism. It emphasizes the need for targeted interventions to support affected communities in adapting to and recovering from environmental crises.

References

Afifi T, Milan A, Etzold B, Schraven B, Rademacher-Schulzb C, Sakdapolrak P, Warner K (2016) Human mobility in response to rainfall variability: opportunities for migration as a successful adaptation strategy in eight case studies. Migr Dev 5(2):254–274

Akhter S, Bauer S (2014) Household level determinants of rural-urban migration in Bangladesh. Int J Soc Hum Sci Eng 8(1):24–27

Bates DC (2002) Environmental refugees? Classifying human migrations caused by environmental change. Popul Environ 23(5):465–477

Bernzen A, Jenkins JC, Braun B (2019) Climate change-induced migration in coastal Bangladesh? A critical assessment of migration drivers in rural households under economic and environmental stress. Geosciences 9(1):51

Black R, Kniveton D, Skeldon R, Coppard D, Murata A, Schmidt-Verkerk K (2008) Demographics and climate change: future trends and their policy implications for migration. Development Research Centre on Migration, Globalisation and Poverty. Brighton: University of Sussex, pp 157–175.

Black R, Adger WN, Arnell NW, Dercon S, Geddes A, Thomas D (2011) The effect of environmental change on human migration. Glob Environ Chang 21:S3–S11

Black R, Arnell NW, Adger WN, Thomas D, Geddes A (2013) Migration, immobility and displacement outcomes following extreme events. Environ Sci Policy 27:S32–S43. https://www.ipcc.ch/report/ar6/wg2/

Bujang MA, Nurakmal B (2016) Sample size guideline for correlation analysis. World J Soc Sci Res 3(1):37–46. https://doi.org/10.22158/wjssr.v3n1p37

Bujang MA (2021) A step-by-step process on sample size determination for medical research. Malay J Med Sci MJMS 28(2):15

Carrico AR, Donato K (2019) Extreme weather and migration: evidence from Bangladesh. Popul Environ 41:1–31

Chen JJ, Mueller V, Jia Y, Tseng SKH (2017) Validating migration responses to flooding using satellite and vital registration data. Am Econ Rev 107(5):441–445

Castles S (2017) International migration at a crossroads. In: The politics of citizenship in immigrant democracies, Routledge, pp 89–106

Dallmann I, Millock K (2017) Climate variability and inter-state migration in India. CESifo Econ Stud 63(4):560–594

Das M, Das A, Momin S, Pandey R (2020) Mapping the effect of climate change on community livelihood vulnerability in the riparian region of Gangatic Plain, India. Ecol Indic 119:106815

Debnath M, Nayak DK (2022) Assessing drought-induced temporary migration as an adaptation strategy: evidence from rural India. Migr Dev 11(3):521–542

Falk GC (2015) Land use change in the coastal regions of Bangladesh: a critical discussion of the impact on delta-morphodynamics, ecology, and society. ASIEN 134(1):47–71

Fussell E (2018) Population displacements and migration patterns in response to Hurricane Katrina. In: Routledge handbook of environmental displacement and migration. Routledge, pp 277–288

Gray CL, Mueller V (2012) Natural disasters and population mobility in Bangladesh. Proc Natl Acad Sci 109(16):6000–6005

Hunter LM, Luna JK, Norton RM (2015) Environmental dimensions of migration. Annu Rev Sociol 41:377–397

Hoffmann R, Wiederkehr C, Dimitrova A, Hermans K (2022) Agricultural livelihoods, adaptation, and environmental migration in sub-Saharan drylands: a meta-analytical review. Environ Res Lett 17(8):083003

IDMC (2021). https://www.internal-displacement.org/global-report/grid2021/

Johnson CA (2018) Climate, migration and displacement: Exploring the politics of preventative action. In: Routledge handbook of environmental displacement and migration. Routledge, pp 342–355

Kniveton D, Schmidt-Verkerk K, Smith C, Black R (2008) Climate change and migration: Improving methodologies to estimate flows. United Nations.

L. Perch-Nielsen S, Bättig M, Imboden D (2008) Exploring the link between climate change and migration. Clim Chang 91(3):375–393

Martin M, Billah M, Siddiqui T, Abrar C, Black R, Kniveton D (2014) Climate-related migration in rural Bangladesh: a behavioural model. Popul Environ 36(1):85–110

McLeman R (2013) Developments in modeling of climate change-related migration. Clim Change 117:599–611

McLeman R (2018) Thresholds in climate migration. Popul Environ 39(4):319–338

McLeman R, Smit B (2006) Migration as an adaptation to climate change. Clim Change 76(1):31–53

McLeman R, Mayo D, Strebeck E, Smit B (2008) Drought adaptation in rural eastern Oklahoma in the 1930s: lessons for climate change adaptation research. Mitig Adapt Strat Glob Change 13:379–400

Meze-Hausken E (2004) Contrasting climate variability and meteorological drought with perceived drought and climate change in northern Ethiopia. Clim Res 27(1):19–31

Mondal M, Biswas A, Haldar S, Mandal S, Bhattacharya S, Paul S (2022) Spatio-temporal behaviours of tropical cyclones over the bay of Bengal Basin in last five decades. Trop Cyclone Res Rev 11(1):1–15

Mondal M, Paul S, Mukhopadhyay A, Mandal P, Chatterjee U, Santra S, Paul S (2023) Evaluating rural poverty and livelihood diversification in the context of climate-induced extreme events in coastal area: Insights from Indian Sundarban. Reg Stud Mar Sci 67:103191

Naujoks D (2021) Trends, Drivers, and Dynamics of Flight and Migration. In: Forced Displacement and Migration: Approaches and Programmes of International Cooperation. Wiesbaden: Springer Fachmedien Wiesbaden, pp 19–40

Neumann K, Hilderink H (2015) Opportunities and challenges for investigating the environment-migration nexus. Human Ecol 43(2):309–322

Panda A (2010) Climate induced migration from Bangladesh to India: Issues and challenges. Available at SSRN 2186397

Renaud FG, Dun O, Warner K, Bogardi J (2011) A decision framework for environmentally induced migration. Int Migr 49:e5–e29

Schoonover R, Cavallo C, Caltabiano I, Femia F, Rezzonico A (2021) The security threat that binds us

Sengar B (2023) From the Sundarban delta to Deccan 'Aurangabad': climatic refugees' resilience for livelihood. J Indian Ocean Reg 19(2):156–166

Stapleton S et al. (2017) Climate change, migration and displacement: the need for a risk-informed and coherent approach, Overseas Development Institute (ODI) and United Nations Development Programme (UNDP), https://www.odi.org/publications/10977-climate-change-migration-and-displacementneed-risk-nformed-and-coherentapproach

Swain A (1996) Displacing the conflict: environmental destruction in Bangladesh and ethnic conflict in India. J Peace Res 33(2):189–204

Swapan MSH, Gavin M (2011) A desert in the delta: participatory assessment of changing livelihoods induced by commercial shrimp farming in Southwest Bangladesh. Ocean Coast Manag 54(1):45–54

Twinomuhangi R, Sseviiri H, Kato AM (2023) Contextualising environmental and climate change migration in Uganda. Local Environ 28(5):580–601

Ullah AKMA (2012) Climate change and climate refugee in Egypt: an overview from policy perspectives. TMC Acad J 7(1):56–70

Viswanathan B, Kumar KK (2015) Weather, agriculture and rural migration: evidence from state and district level migration in India. Environ Dev Econ 20(4): 469–492

Chapter 5
Quantifying Migration Influences: Coastal Perspectives

Abstract Cyclonic hazards significantly impact migration outcomes, influenced by environmental, political, demographic, economic, and social factors. These factors are interconnected with policies that encourage or impede migration. Cyclonic hazards can exert gradual pressure through prolonged inundation or sudden impacts like tropical cyclones. Individuals may migrate in response to immediate threats or anticipate future risks. These factors and their intertwined nature influence the complexity of migration. The study's primary goal is to understand the migration pattern and the socio-economic and livelihood scenarios of migrant and non-migrant families. The whole research was executed based on an empirical data set collected through an intensive field survey through a semi-structured questionnaire. The question was formatted with the help of a pilot survey on those particular areas and taken with the help of existing literature on this field. Initially, we selected ten villages based on their cyclonic risk intensity. Four villages were chosen from high-risk to cyclone, three were selected from moderate-risk areas, and the last three were comparatively less prone to cyclones. Among the total households from the villages, we took 750 households as sample households. Out of the total sample, nearly 65% were categorized as migrated households from where a minimum of one member was migrated, and the remaining 34% were classified as non-migrant families where no one migrated. The data revealed that households near the coast are more prone to cyclones, which have significant damage and are more compelled to migrate. Landholding capacity is the most critical factor in deciding on migration. Landless households or low-holding families were more prone to migrate than those having high landholding capacity. The people who have attended the school up to the lower class or illiterate are more prone to migrate. The people living in Kucha houses are more prone to cyclone damage. Moreover, most of the families from coastal Sundarbans migrated with a dream of making a concrete house. Here, it also established that migration could be an adaptation from the adverse effect of cyclones for coastal dwellers. Most of the people who migrated had better economic conditions after migration, and it was also seen in some cases that migrant households were more economically powerful than non-migrant families. However, it also challenges the govt. Most people work in the destination as an informal job with low social and service security due to unskilled labor. So, governments must take the necessary

P. Mandal and S. Paul, *Climatic Vulnerability and Migration: Evidence from Indian Sundarbans*, Springer Geography, https://doi.org/10.1007/978-3-032-05858-4_5

steps to skill the youth age population for that security concern with "**Migrant as an adaptation.**"

Keywords Cyclone · Landholding capacity · Migrant and non-migrate family · Adaptive capacity · Infrastructure facility · Copping capacity · Women vulnerability

5.1 Introduction

The human decision to migrate is considered a complex behavioral choice guided by the economic, social, demographic, and ecological conditions of the environment within which an individual resides (Hunter et al. 2015; McLeman 2018; Priodarshini and Mallick 2021). Migration has been seen as an adaptive strategy, besides livelihood diversification, in the context of highly adverse environmental conditions or risks (Biswas and Mallick 2021). Migration is described as one of the responses to threats to life, particularly in populated coastal regions prone to climatic hazards (Sengupta and Samanta 2022). Discussions on "climate migration" increasingly recognize the need for studies on actual behavior and a more detailed probabilistic approach. This approach involves analyzing climate stressors alongside other socio-economic and political drivers (Hunter et al. 2015). The International Organization for Migration defines environmental migration as human mobility prompted by sudden or gradual environmental changes that negatively impact people's living conditions, compelling them to leave their usual homes, either voluntarily or involuntarily, within their own countries or abroad, for a temporary or permanent period (McAuliffe and Ruhs 2017). In its fifth assessment report, the Intergovernmental Panel on Climate Change (IPCC) suggests that climate change is a significant factor driving migration, particularly climatic and environmentally induced migration within nations (Twinomuhangi et al. 2023).

Furthermore, Twinomuhangi et al. (2023) emphasize that environmental changes interact with climate change, posing food, economic, and livelihood security risks and fueling migration. Given the critical role of the environment, especially the climate, in sustaining livelihoods, its degradation jeopardizes people's socio-economic well-being, prompting migration (Hollowed et al. 2022). Environmental migration ultimately arises from the sudden onset of extreme climate events and catastrophes such as storms, floods, landslides, and earthquakes, causing displacement and relocation. It also stems from gradual onset processes like rising temperatures, land degradation, and the long-term effects of dry spells and drought on agriculture and ways of life (Twinomuhangi et al. 2023; Mondal et al. 2022).

The intricate relationship between population movement and climate change is embodied in the "Climate mobility nexus." The disruption of multiple environmental factors, including temperature, precipitation, sea levels, and extreme weather events caused by climate change, has significant implications for ecosystems, agriculture, water supplies, and infrastructure (Afifi et al. 2016; Dallmann and Millock 2017;

Viswanathan and Kumar 2015). Climate-induced or environmental migration is triggered by these changes, leading to the relocation of individuals and groups in pursuit of safer living conditions, either within their own countries or across international borders. This underscores the pressing need for comprehensive policy structures that effectively address the underlying causes of climate change and the many issues displaced people face. These initiatives include mitigation methods to reduce greenhouse gas emissions, adaptation techniques to cope with changing environmental circumstances, and policies promoting sustainable development and efficient management of migrant flows. Equally important are the legal protection of displaced people's rights and measures to build resilience, reducing the probability and effects of displacement events. By combining these strategies, societies can aim to respond to the intricate relationships between climate change and human mobility in a more resilient and equitable manner.

According to the 2024 Global Report on Internal Displacement (IDMC 2021), 7.7 million people were internally displaced globally due to disasters at the end of 2023. Although data gaps on the duration of disaster displacement make this figure conservative, it shows that, like conflict, disasters can keep people displaced for long periods. Disasters triggered 26.4 million new internal displacements, or movements, across 148 countries and territories during 2023. This is the third-highest figure in the last decade. A third occurred in China and Türkiye due to severe weather events and high-magnitude earthquakes. Displacements associated with weather-related disasters decreased by a third compared with 2022, partly due to the change from La Niña to El Niño during the year. Storms and floods led to fewer displacements across most of Asia, but floods in other areas triggered record numbers, particularly in the Horn of Africa.

Cyclonic and non-cyclonic migrants are different patterns of human migration, particularly in the context of economic and labor migration. Cyclonic migrants are individuals or groups who migrate suddenly and unpredictably (within 6–12 months of the event), often due to rapid economic changes, political upheaval, or natural disasters. Their movement patterns can be difficult to predict and may change direction quickly, similar to the path of a cyclone. Non-cyclonic migrants, on the other hand, migrate more predictably and steadily, akin to regular weather patterns rather than a sudden storm. They move gradually and incrementally, driven by long-term economic opportunities, educational pursuits, or family reunification. Their migration patterns are more predictable and consistent over time, making it easier for receiving areas to plan and integrate these migrants. Non-cyclonic migrants are more likely to settle permanently or for extended periods in their new locations, contributing to the social and economic fabric of the host communities. They often integrate more smoothly into the local society and economy, such as economic migrants moving to another country for better job opportunities, students migrating for education and employment in the host country, and families moving for long-term stability and improved quality of life. The distinction between cyclonic and non-cyclonic migrants has significant implications for policymakers, planners, and humanitarian organizations. Cyclonic migration requires rapid deployment of resources and emergency aid, while non-cyclonic migration benefits from long-term planning and sustainable development programs. Integration policies for non-cyclonic migrants often focus on language and job training programs, while cyclonic migration may require immediate shelter, healthcare, and security. Understanding these concepts helps develop more effective migration policies and humanitarian responses tailored to the different needs and impacts of various migrant groups.

Indian Sundarban, a vast and dynamic ecosystem within the more significant Sundarban delta, is home to diverse flora and fauna. The region faces various environmental pressures, including severe cyclonic storms, climate change, rising sea levels, and human activities. One notable aspect of this region is the migration of human beings, which can be categorized into cyclonic and non-cyclonic migrants. Cyclonic migrants are species that move or are displaced due to cyclonic activities, such as tropical cyclones, which can significantly impact terrestrial and aquatic ecosystems. Cyclonic migrants can be forced due to habitat destruction, altered food resources, temporary displacement, or other factors. Non-cyclonic migrants, on the other hand, are species that move due to natural seasonal changes, human-induced habitat modifications, or intrinsic biological needs. Seasonal migration is expected in the Sundarban, following rainfall patterns, temperature changes, and breeding cycles. Human-induced habitat changes, such as deforestation, agricultural expansion, and urban development, can lead to habitat fragmentation and degradation, forcing species to relocate. Ecological and biological needs also play a role in migration patterns. Changes in migration patterns can alter the socio-economic balance, potentially leading to the dominance of old age, the decline of women and children, and others. Understanding the dynamics of cyclonic and non-cyclonic migrants in the Indian Sundarban is crucial for developing conservation strategies and mitigating climate changes and human activities' impacts on this unique and fragile ecosystem. Thus, the study aims to explore the patterns of such migrations, the socio-economic and environmental drivers behind them, and their implications for both the source and destination areas. By focusing on the Indian Sundarbans, the research highlights the critical need for integrated climate resilience and sustainable development policies to address the challenges posed by climatic migration.

5.2 Background of the Study

The data on climatic displacement in India from 2008 to 2023 provides a profound insight into the trends and impacts of natural hazards on population movements' data (https://www.internal-displacement.org/database/displacement-data/). According to the Internal Displacement Monitoring Centre dataset, there is an average displacement of around 170,000 individuals per event, with a median displacement of approximately 1,500 people. The most extreme recorded displacement event affected up to 6.9 million individuals. Among the various types of natural hazards, floods emerge as the most frequent and consequential, often leading to displacements of over 2 million people in some instances. The severity of storms, particularly cyclones, is evident

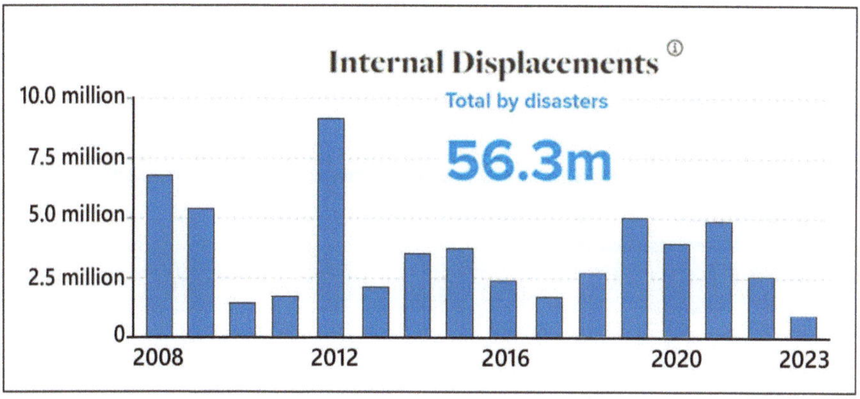

Fig. 5.1 Temporal scenario of Internal Displacement due to disaster in India from 2008 to 2023 (*Source* Internal Displacement Monitoring Centre (IDMC))

as they precipitate significant population movements. These events underscore such natural phenomena' extreme intensity and destructive potential.

As depicted in Fig. 5.1, these cyclonic events are a critical factor in the displacement data, highlighting the urgent need for robust disaster management strategies. Droughts, albeit less frequent, contribute notably to displacement figures. These events reflect prolonged environmental and economic stresses that cumulatively lead to significant population dislocations. Despite their lower frequency, the impact of droughts cannot be underestimated, as they highlight the chronic nature of specific climatic challenges faced by India. Over the years, there has been a discernible increase in the number of displacements, suggesting a rise in the frequency of climatic events, improvements in reporting mechanisms, or a combination of both factors. The data exhibits substantial variability in displacement figures, with some years experiencing more severe or significant events leading to large-scale displacements.

This trend underscores the escalating nature of climatic threats and the need for continuous monitoring and adaptation in disaster response strategies. Floods consistently remain the predominant cause of displacement in India. Their recurrent nature and capacity to cause massive population movements necessitate implementing effective early warning systems and comprehensive disaster preparedness plans. The significant displacement caused by storms further emphasizes this requirement.

The data indicates that despite their infrequency, droughts still contribute substantially to the overall displacement figures, illustrating certain climatic conditions' persistent and chronic impacts. The analysis of this data underscores the imperative for sustained investment in disaster preparedness and the development of resilient infrastructure. As India grapples with increasing frequency and intensity of climatic events, the need for robust and adaptive strategies becomes more pressing. The insights provided by this dataset (illustrated further in Fig. 5.2) call for a multifaceted approach to mitigate the adverse effects of natural hazards on populations. This involves enhancing early warning systems and disaster response capabilities and

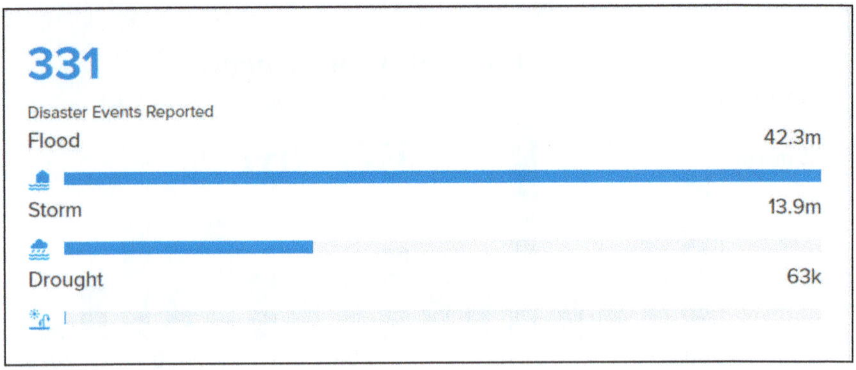

Fig. 5.2 The status of internal displacement due to different disasters in India from 2008 to 2023 (*Source* Internal Displace Monitoring Center (IDMC))

fostering long-term resilience through infrastructural development and sustainable environmental management practices. The climatic displacement data from India reveals significant trends and impacts of natural hazards, with floods and storms being the most prominent causes of displacement. These events' increasing frequency and intensity underscore the necessity for comprehensive disaster preparedness and resilient infrastructure. The data calls for a concerted effort to enhance early warning systems, improve reporting mechanisms, and invest in sustainable solutions to mitigate climatic events' chronic and acute impacts on India's population.

5.3 Materials and Methods

5.3.1 Study Area

The Pathar Pratima CD Block in West Bengal, India, is home to more than 0.33 million people and faces frequent hydro-geomorphologic changes due to cyclones and climate change issues. Nearly half of households are below the poverty line, with the Sundarban region having the highest poverty rates. The region's workforce relies heavily on physical capital for livelihoods, with 75% relying on it. The coastal areas are highly susceptible to cyclones, exacerbating living conditions and livelihoods. Recent cyclones highlight the need for effective disaster management and resilience-building measures.

5.3.2 Selection of the Variables

The section explores the multifaceted dynamics of male out-migration in rural areas, emphasizing the central role of landholding size as a primary economic factor influencing migration decisions. This notion is supported by various studies (Parganiha and Sharma 2017), highlighting how land ownership status affects migration patterns. Research indicates that temporary or seasonal male out-migration often correlates with job availability in destination areas (Keshri and Bhagat 2013). Landless households, lacking stable income sources, frequently migrate to attain a better standard of living (Bernzen et al. 2019). However, during periods of drought, the intensity of out-migration significantly escalates, often becoming a last resort for affected individuals (Bhagat et al. 2018). Indebtedness is another crucial factor driving seasonal out-migration, irrespective of landholding size (Debnath and Nayak 2022).

Moreover, a strong correlation exists between landholding size and the duration of out-migration (Debnath and Nayak 2022), underscoring the intricate relationship between economic status and migration dynamics. Social networks and informal connections are pivotal in facilitating migration decisions, with Information from friends deemed the most influential. Intermediaries and relatives also serve as essential sources of Information, further shaping individuals' migration choices. The working-age group of males emerges as a significant parameter influencing the intensity of male out-migration (Debnath and Nayak 2022), highlighting the demographic aspect of migration trends. Importantly, research by Debnath and Nayak (2022) reveals that a considerable proportion of scheduled castes and tribes are landless, indicating a higher incidence of out-migration among these marginalized groups. The paragraph elucidates the intricate interplay between economic, social, and demographic factors in shaping male out-migration in rural areas. It underscores the significance of landholding size, job opportunities, indebtedness, social networks, and demographic characteristics in understanding migration patterns and highlights the disproportionate impact of migration on marginalized communities (Table 5.1).

5.3.3 Methods

The present research aims to investigate the impact of cyclone-induced migration among coastal inhabitants of the Indian Sundarbans. Data for this study were gathered through a household survey, which aimed to capture Information regarding the scale, nature, and process of rural out-migration. A structured questionnaire was utilized for this purpose. To evaluate cyclone-induced migration at the household level, households were categorized into two main groups: migrated and non-migrated. Migrated households were defined as those from which at least one individual had migrated, while non-migrated households were those from which no members had migrated. Determinant variables were selected and categorized accordingly. The first predictor variable considered was the village's location, categorized into three groups: coastal

Table 5.1 Explain the selected variables

Variables		Sub-categories	Base (0)	Scale	Source
Demographic aspect	Family size	(0–3)	(0–3)	Number (household)	Paul et al. (2022)
		(4–6)			
		(6–9)			
	Social categories	UR	UR	Number (household)	
		SC			
		OBC-B			
		OBC-A			
	Religion	Hindu	Hindu	Number (household)	
		Muslim			
	Family type	Joint	Joint	Number (household)	
		Nuclear			
Economic and coping capacity	Number of earning members	One earning members	One earning member	Number (person)	Debnath and Nayak (2022)
		Two earning members			
		Three earning members			
	Income before migration	<250	<250	Number (household)	
		250–500			
		>500			
	House structure before Amphan	Muddy	Muddy	Number (household)	
		Semi-cemented			
		Fully cemented			
	House structure after Amphan	Muddy	Muddy	Number (household)	
		Semi-cemented			
		Fully cemented			
	Survive post-cyclonic migration	By saving	Others	Number (household)	
		Others			
Social capital	Relief was added after the cyclone	Yes	Yes	Number (household)	Naser et al. (2019)
		No			
	Borrow money	Yes	Yes	Number (household)	
		No			
	How did you select the destination?	Friends help	By the contractor		
		Had prior information			
		Relative lived there			

(continued)

Table 5.1 (continued)

Variables		Sub-categories	Base (0)	Scale	Source
Livelihood crisis	Damaged roof	Fully damaged	Fully damaged	Number (household)	Benrzen et al. (2019)
		Partially damaged			
		No damaged			
	Crop damaged	2–5 quintal	2 to 5 quintals	Quintal	
		5–10 quintal			
		>10 quintal			
	Fishing cultivation damaged in rupees	10,000 to 20,000 damaged	10,000 to 20,000 damaged	INR	
		20,000 to 30,000 damaged			
		>30,000 damaged			
Exposure and risk to cyclone	Location of villages	Riverine	Inland	Location	Mondal et al. (2021)
		Coastal			
		Inland			
	Duration of post-cyclone disturbances	Up to 1 week	Up to 1 week	Time	
		Up to 2 weeks			
		Up to 3 weeks			
		>1 Month			
Natural capital	Landholding capacity	Landless	Landless	Bigha	Debnath and Nayak (2022)
		Marginal landholder			
		Small landholder			

Source Developed by the researcher

villages near the coast, riverine villages near rivers, and inland villages surrounded by land.

The social status of households was classified into four groups: General, Scheduled Caste (SC), Other Backward Class-A (OBC-A), and Other Backward Class-B (OBC-B). Religion was categorized into Hindu and Muslim groups. Furthermore, the family structure was divided into joint and nuclear families. Family size was categorized into three groups: households with 1–3 members, households with 4–6 members, and households with more than seven members. Lastly, the number of earning members within each household was grouped into two categories: households with one earning member and households with two or three earning members. This comprehensive categorization allows for a nuanced analysis of the factors influencing cyclone-induced migration among coastal communities, considering variables such as geographical location, social status, religious affiliation, family structure, family size, and the number of earning members within households. By considering these factors, the research aims to provide valuable insights into the complexities of cyclone-induced migration dynamics in the Sundarbans region.

The methodology assessed cyclonic vulnerability and its impact on out-migration in cyclone-prone areas. One crucial variable under consideration is the distance between households and the coastline or river bank, categorized into three groups: within 500 m, 500 m to 1 km, and more than 1 km. This categorization allows for examining how proximity to water bodies influences cyclone vulnerability. Furthermore, the structural integrity of houses before and after migration is evaluated, with classifications including hut or muddy, cemented, and fully cemented structures. The extent of damage to dwelling houses post-cyclone is categorized into fully, partially, and undamaged, providing insight into the severity of cyclonic impact on housing infrastructure.

Similarly, the damage to crops and fishing cultivation due to cyclones is assessed by categorizing the extent of damage into three groups based on quintal or monetary value thresholds. The categories allow for understanding the economic losses incurred by households engaged in agriculture and fishing activities. The duration of environmental disruption following a cyclone is also considered, with categories ranging from one week to one month or more. So, the duration of adverse environmental conditions affects livelihoods and living conditions. Additionally, the landholding capacity of households is classified into three categories: landless households with no land for cultivation, marginal families with up to 2 Bigha of land, and small landholders with more than 2 Bigha of cultivated land. This classification helps assess the relationship between land ownership and vulnerability to cyclones.

5.3.4 Statistical Model

A binary logistic regression model is employed to analyze the impact of these determinant factors on out-migration. The dependent variable is coded as "1" if the household migrated and "0" if it did not. This statistical approach identifies significant factors influencing migration decisions in cyclone-prone areas, providing valuable insights for disaster management and mitigation strategies.

The logistic model can be expressed as follows.

$$p(MS_i = j) = \frac{e^{\sum \gamma_i x_i}}{1 + e^{\sum \gamma_i x_i}}, \quad \text{for } j = 0, 1 \tag{5.1}$$

The equation used to estimate the coefficients is as the form,

$$\log \frac{p_i}{1 - p_i} = \beta_0 + \beta_1 x_1 \cdots + \beta_m x_m, \tag{5.2}$$

where $\beta_0, \ldots, \beta_M$ are regression coefficients indicating the relative effect of the particular independent variable on the dependent variable? The ratio $\left(\frac{P_i}{1-P_i}\right)$ is called odds, the probability ratios of institutional and non-institutional delivery status.

Based on the results of this inquiry, related studies, and existing literature, the independent variables chosen are as follows. The independent variable categories were coded during the logistic run. Families 1 through 3 are coded as "1," families 4 through 6 as "2," and families more significant than six as "3." There are 500 m with code "1," 500 m to 1 km with code "2," and more than 1 km with code "3." Up to five quintals of damaged crop are coded as "1," five to ten quintals as "2," and more than ten quintals as "3." The codes for fishing cultivation damages were "0," "1," "2," or "3," denoting no damage, up to 10,000 rupees, 10,000 to 30,000, and over 30,000, in that order.

5.3.5 *Model Accuracy*

Before proceeding with the model fitting process, we conducted a variance inflation factor (VIF) analysis to examine the potential presence of Multicollinearity among the independent variables. Multicollinearity occurs when independent variables in a regression model are highly correlated, which can lead to unreliable and unstable estimates of the coefficients. In our study, all variables had VIF scores of 2 or less, indicating no significant Multicollinearity among the independent variables. Typically, a VIF exceeding five indicates Multicollinearity (Daoud 2016). Since our variables had VIF scores below this threshold, we included all essential variables in the model without concern for Multicollinearity issues. This step was necessary to ensure the reliability and validity of our regression model, as Multicollinearity can lead to biased estimates and incorrect inferences about the relationships between the independent variables and the dependent variable.

For optimal classification, the cut-off value we took that best suits our assumptions, 0.5, is derived from the sensitivity and specificity curve, which is closer to the point of intersection of the sensitivity and specificity curve (Fig. 5.3). Among the 147 observations out of 60, our model accurately classifies 50 instances. In the case of the 87 observed instances with no outcomes, our model correctly identifies 76, resulting in an impressive overall accuracy of 85.2% (Table 5.2).

The area under the Receiver Operating Characteristic (ROC) curve, as shown in Fig. 5.3, provides additional evidence of the model's capability. This performance metric highlights the reliability and effectiveness of our model in classification tasks. With an area under the curve of 0.949 (Table 5.3), our model exhibits strong performance compared to the dataset. This metric indicates that the model can effectively distinguish between different classes, demonstrating its overall competence and suitability for the classification work (Fig. 5.3).

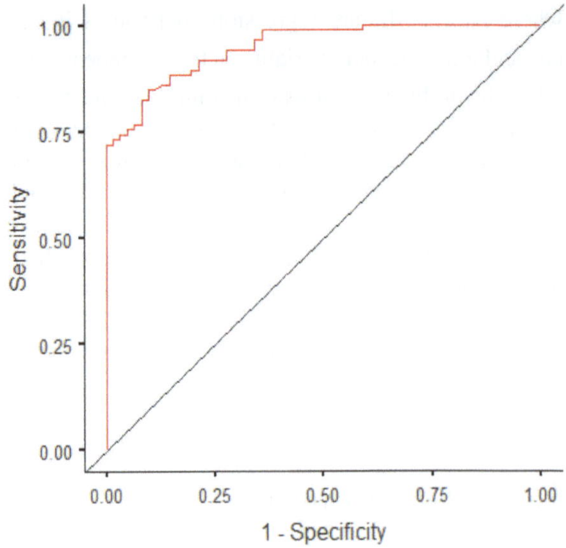

Fig. 5.3 Validation of the model by ROC curve

Table 5.2 Overall model accuracy of logistic regression

Model fit measures						
Model	Deviance	AIC	R^2_{McF}	Overall model test		
				χ^2	df	p
1	80.6	147	0.596	119	32	<0.001

Source Computed by the researcher

Table 5.3 Performance metrics of predictive model

Predictive measures			
Accuracy	Specificity	Sensitivity	AUC
0.857	0.820	0.884	0.949

Note The cut-off value is set to 0.5

5.4 Results

5.4.1 Determinants of the Cyclonic-Induced Migration

The study aims to understand the pattern and influencing factors that decide to out-migrate from a multi-hazard zone like (Fig. 5.4) the Indian coastal Sundarban. To fulfill the objective initially, we have selected nine villages in proximity to the coastal line, three villages near the coast, three from the riverfront, and the rest of the Inland villages selected. A previous study (Mondal et al. 2022) established that the intensity of cyclonic vulnerability varies from coast to Inland, as the intensity of cyclones

Fig. 5.4 Vegetation and Infrastructure was lost by cyclone Amphan. *Source* By authors on 09-05-2023

is higher in coastal villages, which is less for riverfront areas but higher than in Inland areas. It was also established by an existing study (Bernzen et al. 2019) that the out-migration of Male workers varies according to the intensity of cyclonic exposure and damage to their livelihood condition. Besides the cyclonic exposure, the decision to migrate depends on socio-economic and demographic variables. Here, we have applied the logistic regression to determine the determinant factors of rural out-migration. Here, we have taken Household type as a dependent variable.

Two types of households have been categorized: a migrated household and the other is a non-migrated household. If one member or more from a family goes outside to do work, it is treated as a Migrant household, coded as "1," and on the other side, no one migrated from a family as a non-migrant household, coded as "0." The independent variables have been selected based on the present investigation, similar past studies, and the existing literature. Before discussing the determinant factors of migration, describing the descriptive or overall scenarios of independent variables will be suitable for understanding the overall situation. For the study, 750 households were surveyed from these selected villages with a 5% confidence level (Eq. 5.1). Among them, 487 households are under-migrated households, with 65%, and the rest are treated as non-migrated households, with 35% (263). The geographical location (Coastal, riverine, and inland) wise distributions of households are 40%, 25%, and 35% for Coastal, riverine, and inland villages, respectively.

5.4.2 Determinants of Demographic Conditions

The logistic regression analysis provides insightful findings regarding the vulnerability of different demographic characteristics of a household to cyclone-induced migration. Social dynamics play a pivotal role in migration susceptibility. Notably, households in the Scheduled Caste (SC) category exhibit a 3.4 times higher propensity for cyclone-induced migration than General category households. However, religion can influence migration from coastal Sundarban, which needs to be researched in the future. Family structure emerges as another influential factor. Nuclear families demonstrate a lower likelihood of migration, yet they are approximately 77% less likely to migrate than joint families in such circumstances.

Furthermore, the analysis delves into the impact of family size and composition on migration tendencies. Larger families face significantly high migration risks, with households of greater size being 7.57 times more likely to migrate than smaller ones. Specifically, households with more than three male members exhibit a nearly twofold increase in the likelihood of migration compared to those with fewer than three male members. For instance, the likelihood of migration during cyclones differs significantly based on the number of female members within a household.

A household with fewer female members exhibits a notably higher migration risk. Those with low female membership (Fig. 5.5) have 87%-time odds to migrate during cyclonic years than families with 4–6 female members (Fig. 5.6). The result also shows that they are less likely to migrate with an increased education level. Educational attainment emerges as another critical determinant. Individuals who completed schooling up to the 12th grade demonstrate a 19% times lower likelihood of migration during cyclones than those with less than 5 years of schooling.

Conversely, the probability of migration increases substantially for individuals with 5 years of schooling, reaching 92% times higher likelihood compared to those with schooling beyond the 12th grade. Notably, there are no significant findings regarding migration likelihood for individuals with 5 or 10 years of schooling. The odds ratio reveals that the people migrating from this region are 4.78 times more likely to migrate temporarily than permanently. These findings underscore the multifaceted

Fig. 5.5 The young man stated that he had migrated before Amphan, but after that he didn't migrate due to the absence of any male member in his family, and he is selling fish. So, security is a matter to make the decision to migration. *Source* By authors on 18-11-2023

Fig. 5.6 Women share their struggles in a climatically vulnerable region where their husbands are migrant workers. *Source* By authors on 18-11-2023

nature of migration, elucidating the differential vulnerabilities across geographical, social, and familial dimensions. Such insights are invaluable for informing targeted interventions and policy measures to mitigate the adverse impacts of cyclones on vulnerable populations (Table 5.4).

5.4.3 Determinants of the Economic and Coping Phenomenon of Migration

The analysis of migration patterns during cyclonic events reveals a complex interplay of demographic, socio-economic, and environmental factors. Notably, family size and the number of earning members are significant determinants. Larger families, consisting of 2–3 members and families with 2–3 earning members, have 1.21- and 38%-times higher odds of migrating, respectively, compared to smaller families with only one member or one earning member populations (Table 5.5). So, the economic pressure and resource allocation within larger families compel them to seek better opportunities through migration. Economic factors further elucidate the migration propensity. Families with daily earnings exceeding 500 rupees exhibit a 14% lower likelihood of migration (Fig. 5.7). So, higher daily earnings buffer against the economic shocks precipitated by cyclonic events. Conversely, households with lower-income members face an 86% higher likelihood of migrating than their higher-income counterparts. This stark contrast underscores the vulnerability of economically disadvantaged families, who lack the financial resilience to withstand the impacts of such disasters.

Table 5.4 Determinants of demographic characteristics to migration

Predictor	Estimate	SE	Z	Odds ratio
Categories (SC/ST/OBC-A/OBC-B/GEN)				
SC-UR	1.49	0.535	2.78	4.427**
OBC-A-UR	1.61	0.705	2.28	4.981**
Religion (Hindu/Muslim/others)				
Muslim-Hindu	0.107	0.798	0.134	1.113
Household structure: joint/nuclear				
Nuclear-jointly	−1.453	0.473	−3.07	0.234**
No. of family member:1: (1–3), 2: (4–6), 3: (7–9)				
2–1	1.62	0.69	2.34	5.04**
3–1	2.15	0.795	2.7	8.57**
Male: 1: (1–2), 2: (3–4), 3: (4–6)				
2–1	1.084	0.463	2.34	2.95**
Female 1: (1–2), 2: (3–4), 3: (4–6)				
2–1	−4.277	2.6	−1.6428	0.38
3–1	−5.055	4.59	−1.1015	0.13**
Education status of migrant worker: 1—0 to 5 years, 2—6 to 10 years, 3—11 to 15 years, 4—>15 years education				
2–1	−0.657	0.542	−1.212	0.52
3–1	−1.658	0.699	−2.374	0.19**
4–1	−2.485	1.136	−2.186	0.08**
What type of migration was it? 1. Permanent, 2. Temporary				
2–1	1.76	0.513	3.42	5.786***

Note ** $p < 0.01$; *** $p < 0.001$ (*Source* Computed by the researcher based on primary survey)

The type of housing significantly influences migration decisions. Individuals residing in muddy houses are 87% more likely to migrate during cyclones compared to those in cemented or Pucca houses. This finding highlights the critical role of housing quality in determining the capacity to endure cyclonic conditions. The lack of a significant relationship between muddy and semi-cemented households suggests that while both are vulnerable, there may be marginal differences in their susceptibility. Savings also play a pivotal role in migration decisions. Individuals who rely on their savings for livelihood exhibit a 28% lower chance of migrating than those without savings. This finding reveals that savings serve as a crucial safety net, reducing the need for displacement by providing financial stability during crises. The logistic regression results indicate a 72% higher likelihood of migration among those without savings, emphasizing the critical role of financial preparedness in mitigating migration risk populations (Table 5.5).

Table 5.5 Determinants of economic situation and also coping capacity to migration

Predictor	Estimate	SE	Z	Odds ratio
Number of earning members				
2–1	0.793	0.485	1.635	2.21
3–1	0.323	0.723	0.447	1.381**
Before Amphan, Daily income: 1—<250, 2—250–500, 3—>500				
2–1	−1.925	0.559	−3.44	0.146***
3–1	−1.956	0.734	−2.66	0.141**
Before Amphan house structure: 1—Muddy, 2—Semi-cemented, 3—Cemented				
2–1	−0.89	0.543	−1.639	0.411
3–1	−2	0.814	−2.457	0.135**
How to survive in the post-cyclone situation: By saving money—1, Other—0				
1–0	−1.269	0.467	−2.719	0.281**

Note ** $p < 0.01$; *** $p < 0.001$
Source Computed by the researcher based on primary survey

Fig. 5.7 In Sundarbans, cyclones can't break their spirit; only poverty can. *Source* By authors on 18-11-2023

These findings underscore the intricate nexus of demographic, socio-economic, and environmental factors shaping cyclone-induced migration patterns. The heightened vulnerability of larger families, lower-income households, and individuals living in substandard housing highlights the urgent need for targeted interventions. Enhancing economic resilience through increased earnings, improving housing quality, and promoting savings are essential strategies to support vulnerable populations during cyclonic events. The study illuminates the multifaceted determinants of migration during cyclones, emphasizing the importance of a comprehensive approach to disaster management. Addressing the underlying vulnerabilities related to family size, economic status, housing quality, and savings can significantly reduce migration propensity and enhance the resilience of at-risk communities. The insights from this analysis underscore the need for policies and interventions that prioritize the most vulnerable groups, ensuring their stability and safety during environmental crises.

5.4.4 Social Capital Influences the Migration

The dynamics of cyclone-induced migration are complex, shaped by various factors such as social support, housing conditions, financial resilience, and the availability of relief aid. Those migrating with the assistance of friends exhibit a 23% lower likelihood of migrating during cyclonic years compared to those relying on contractors. So, social networks play a significant role in providing support and reducing the necessity for migration. Friends offer immediate, trustworthy, and non-commercial support, which can mitigate the urgency to relocate during crises. Housing dynamics play a pivotal role in migration decisions. The quality of housing often dictates the extent of damage sustained during cyclones, influencing the inhabitants' decision to stay or migrate. Crop losses further exacerbate the situation, particularly for agrarian communities whose livelihoods depend on agricultural output. The compounded effect of housing damage and crop loss can drive large-scale migration as families seek more stable living conditions and alternative income sources. Relief aid from government and non-governmental organizations (NGOs) significantly impacts post-cyclone migration decisions.

The data reveals that individuals who received relief aid post-cyclone were 23% less likely to migrate immediately after the event. This underscores the importance of timely and adequate relief measures to stabilize affected populations. Conversely, those who did not receive relief aid were 77% more prone to migrate, highlighting the desperation and lack of resources to cope with the immediate aftermath of cyclonic damage (Table 5.6). Financial resilience, or the lack thereof, critically influences migration patterns. Individuals who borrow money from relatives or others are 2.13 times more likely to migrate than those who do not take loans. This finding indicates that borrowing, often seen as a last resort, imposes an additional financial burden that can compel individuals to migrate in search of better economic opportunities to repay their debts and rebuild their homes. Many are forced to leave their familiar surroundings to repay loans and rebuild destroyed residences, underscoring the economic scarcity experienced by vulnerable groups.

Understanding these multifaceted factors is essential for devising targeted interventions to mitigate the adverse impacts of cyclones on vulnerable populations. Effective disaster management strategies should prioritize enhancing social support networks, improving housing resilience, providing timely and adequate relief aid, and

Table 5.6 Social capital

Predictor	Estimate	SE	Z	Odds ratio
Did you get any relief ads from the government after the cyclone? and NGO: 1—Yes, 0—No				
1–0	−1.459	0.536	−2.72	0.233**
By borrowing money from relatives				
1–0	1.14	0.467	2.45	3.136**

Note ** $p < 0.01$; *** $p < 0.001$ (*Source* Computed by the researcher based on primary survey)

bolstering financial stability. Strengthening these areas can significantly reduce the propensity for forced migration and help stabilize affected communities in the aftermath of cyclonic events. The intricate interplay of social support, housing quality, financial resilience, and relief aid significantly shapes cyclone-induced migration patterns. Targeted interventions that address these factors can mitigate the adverse impacts on vulnerable populations, fostering greater resilience and stability in the face of environmental disasters. The insights from this analysis underscore the critical need for comprehensive and multifaceted approaches to disaster management and relief efforts (Table 5.6).

5.4.5 Livelihood Crisis

Field surveys and extensive literature establish that a significant portion of the population from the Sundarbans region migrates to other states to construct a permanent, or "pucca," house for them. The survey results indicate that households experiencing destruction of their homes due to cyclones (Fig. 5.8) are 44% less likely to migrate than those with partially damaged homes, with a 95% confidence level. This suggests that families with fully damaged homes may lack the resources or immediate capacity to migrate due to the sheer extent of their losses and the need to focus on immediate recovery. Intriguingly, households with no damage are 2.4 times more likely to migrate than those with fully damaged homes. This counterintuitive finding could be explained by the more excellent financial stability and mobility of households that sustain no damage, enabling them to pursue better opportunities elsewhere without the immediate burden of reconstruction. Crop loss, categorized into three distinct levels—less than five quintals, 5 to 10 quintals, and more than ten quintals—was analyzed using logistic regression to understand its impact on migration. The analysis revealed no significant difference between the first two categories (less than five quintals and 5 to 10 quintals) regarding their influence on family migration decisions.

However, a significant difference exists when comparing losses of less than five quintals to those exceeding ten quintals. Interestingly, moderate crop damage (5–10 quintals) reduces the odds of migration compared to lower damage (less than five quintals) by a factor of 0.103. This suggests that families facing moderate crop

Fig. 5.8 Cyclone Amphan totally wrecked the family's house. Since the government offered them neither material nor financial support, they were able to build a makeshift thatched house using their own limited resources in the aftermath. For basic shelter repair, they were obliged to rely entirely on their own work and finances due to the absence of institutional support. *Source* By authors on 18-11-2023

damage may still have sufficient resources to sustain themselves and are, therefore, less inclined to migrate compared to those experiencing very low or very high levels of crop loss (Fig. 5.9). The impact of damage to fishing cultivation was also scrutinized, with damages categorized as no damage, 10,000 to 30,000 rupees, and more than 30,000 rupees (Fig. 5.10). The results indicate no significant difference among these categories except for damages between 10,000 and 20,000 rupees, which increases the odds of migration by 3.57 times odds compared to no damage. This suggests a threshold effect where moderate financial losses in fishing cultivation compel households to seek livelihoods elsewhere.

Furthermore, the duration of the post-cyclone effects plays a crucial role in migration decisions. In areas where the after-effects of the Cyclone linger longer, there is a higher probability of migration. This highlights these households' prolonged struggle and instability, pushing them to seek more stable environments. These findings elucidate the complex interplay of housing damage, crop loss, and financial setbacks in influencing migration patterns. They underscore the need for targeted interventions that address both the immediate and long-term needs of affected populations, such as improving housing resilience, providing timely financial aid, and supporting sustainable livelihood options to reduce the necessity of migration (Table 5.7).

Fig. 5.9 The agricultural field behind her home was farmed in the 2019 season, according to a village woman. But Cyclone Bulbul hit the area right before harvest, bringing with it flooding and a lot of rain, which wiped off the mature crops. They had put a lot of work and money into the cultivation, so this abrupt loss of produce not only affected the household's food supply but also caused them to suffer greatly financially. *Source* By authors on 18-11-2023

Fig. 5.10 Their principal source of income is fishing, which both the husband and wife participate. They live close to the coast and frequently go out to sea to collect fish, where they frequently face hazardous circumstances and erratic weather. Their reliance on marine resources emphasizes the vulnerability and economic significance of coastal livelihoods. *Source* By authors on 18-11-2023

Table 5.7 Livelihood crisis and damage due to Cyclone

Predictor	Estimate	SE	Z	Odds ratio
Damaged roof: 1—fully damaged, 2—partially damaged, 3—no damage				
2–1	−0.817	0.374	−2.19	0.442**
3–1	1.229	0.605	2.03	3.419**
Crop damage (production in quintal): 1—2 to 5, 2—5 to 10, 3—>10 quintal				
2–1	−0.237	0.446	−0.532	0.789
3–1	−2.27	1.097	−2.069	0.103**
Fishing cultivation damage (INR) 1—10,000 to 20,000, 2—20,000 to 30,000, and 3—more than 30,000				
2–1	0.9445	0.704	1.341	2.57
3–1	2.1484	1.073	2.0024	3.57**

Note ** $p < 0.01$; *** $p < 0.001$ (*Source* Computed by the researcher based on primary survey)

5.4.6 Exposure and Risk to Cyclonic Migration

The model coefficient table (Table 5.8) reveals substantial migration pattern differences among coastal, riverine, and inland village households (Fig. 5.11). The logistic regression analysis identifies several critical factors influencing household migration, with statistically significant results ($p < 0.05$), providing robust insights into these dynamics. Households residing along riverfronts exhibit a significantly heightened migration risk, being 2.52 times more likely to migrate than those in inland areas. This increased risk can be attributed to the frequent flooding and erosion riverine communities face, which disrupts their livelihoods and compels them to seek more stable environments. The migration risk amplifies dramatically for coastal households (Fig. 5.12), who are 5.5 times more likely to migrate than their inland counterparts. Coastal areas are particularly vulnerable to cyclonic storms and rising sea levels, which lead to severe property damage and loss of livelihood, thereby driving higher migration rates. The proximity of households to rivers or coastlines also influences migration decisions. Living farther from these water bodies reduces the odds of migration, with households located 500 m to 1 km away having an odds ratio of 0.529 compared to those within 500 m. However, migration decisions between households located within 500 m and those between 500 m and 1 km are the same. This suggests that the immediate vicinity of water bodies poses a consistent risk, while the mitigating effect of distance becomes more pronounced beyond the 1-km mark.

The duration of the post-cyclone effects is another crucial factor influencing migration. Households experiencing prolonged post-cyclone impacts show a significantly higher probability of migrating. The odds of migration increase with the length of post-cyclone effects, with an odds ratio of 2.692 for 2 weeks, 3.694 for 3 weeks, and a staggering 5.25 for 4 weeks compared to those experiencing effects

Table 5.8 Exposure and risk to cyclone

Predictor	Estimate	SE	Z	Odds ratio
Location of village: riverine/coastal/inland				
Riverine–inland	1.44	0.575	2.5	2.52**
Coastal–inland	1.89	0.814	2.32	6.6**
Distance to river/coastline: 1—<500 m, 2—500 to 1 km, 3—more than 1 km				
2–1	−0.711	0.528	−1.348	0.491
3–1	−1.935	0.702	−2.756	0.144**
Duration of post-cyclone disturbances: 1—up to one week, 2—one to two weeks, 3—up to three weeks, and four more than one month				
2–1	0.99	0.441	2.25	2.692
3–1	1.307	0.543	2.41	3.694**
4–1	1.658	0.631	2.63	5.25**

Note ** $p < 0.01$; *** $p < 0.001$ (*Source* Computed by the researcher based on primary survey)

Fig. 5.11 Compared to coastal regions, inland places are relatively less vulnerable to disaster. Inland areas suffer much less environmental vulnerability and physical destruction than coastal settlements, which are often subject to tidal surges, high-intensity cyclonic storms, and saline water intrusion. The population's movement patterns and stability of livelihood are greatly impacted by this disparity in exposure. *Source* By authors on 18-11-2023

for just 1 week. This highlight affected households' prolonged struggle and instability, exacerbating their need to relocate to more secure and stable environments. The findings from the logistic regression analysis underscore the profound impact of geographical location and post-cyclone recovery duration on migration decisions.

Fig. 5.12 Coastal bank was broken by Cyclone Amphan *Source* By authors on 18-11-2023

Riverine and coastal households face significantly higher migration risks due to their vulnerability to flooding, erosion, and cyclonic storms.

Additionally, the extended duration of post-cyclone effects further compels migration, reflecting the ongoing challenges and instability these households endure. These insights highlight the necessity for targeted interventions that address the specific needs of riverine and coastal communities, such as improved flood defenses, resilient infrastructure, and timely disaster relief. Enhancing the resilience of these vulnerable populations can mitigate the need for migration and promote stability in the face of increasing climatic threats. The findings call for comprehensive disaster management strategies incorporating immediate relief and long-term resilience-building measures to support at-risk communities (Table 5.8) effectively.

5.4.7 Natural Capital Influences on the Migration

Numerous scholars have established that landholding capacity is a critical determinant of rural out-migration. Small landholders are 0.347 times less likely to migrate than landless households, implying that landless households have a 65% higher propensity to migrate than those owning more than two Bigha of land. This significant difference underscores the pivotal role of land ownership in providing economic stability and reducing the necessity for migration. Interestingly, the analysis reveals no significant difference in migration likelihood between marginal landholders (owning less than one Bigha) and small landholders. This suggests that more land ownership is needed to mitigate the pressures driving migration substantially.

Table 5.9 Natural capital

Predictor	Estimate	SE	Z	Odds ratio
Landholding capacity				
Marginal landholder–Landless	−0.581	0.366	−1.58	0.35
Small landholder–Landless	−1.058	0.535	−1.98	0.56**

Note ** $p < 0.01$; *** $p < 0.001$ (*Source* Computed by the researcher based on primary survey)

The model's predictive accuracy is notable, correctly identifying migrating households 88.4% of the time and non-migrating households 82% of the time. These high accuracy rates underscore the robustness of the model in discerning the factors influencing migration decisions, particularly the critical impact of landholding capacity. This analysis highlights the need for policies that enhance land tenure security and support for landless and marginal landholding households to reduce forced migration and improve rural livelihoods (Table 5.9).

5.4.8 Cyclone Compelled to Migration

In this study, we sought to evaluate the impact of cyclones on migration from the coastal Sundarbans region using the difference-in-difference (DiD) methodology. We focused on two independent variables: the change in household building structure and the distance of the house from the coast or river. The change in household structure was assessed by comparing the condition of houses before and after the Cyclone, resulting in the variable "Change." Subsequently, we created an explanatory variable representing the interaction between changes in housing structure and distance from the coast or river. This analysis generated four distinct conditions: no change within 1 km, no change beyond 1 km, change within 1 km, and change beyond 1 km. We then binary-coded these conditions, assigning a value of 1 to the condition of change within 1 km, while the other three conditions were coded as 0. Employing logistic regression allowed us to assess the effects of cyclonic events on migration decisions.

The results indicate that cyclonic vulnerability influences migration decisions both directly and indirectly. Specifically, the logistic regression analysis reveals that individuals residing within 1 km of the coast or river who experienced property loss are 34% more likely to migrate than those living beyond 1 km who did not suffer property damage. This finding underscores the heightened vulnerability of households near water bodies, where the risk of property loss due to cyclones is significantly higher. The interaction between housing structure changes and distance from the coast or river is crucial in understanding the migration dynamics in the Sundarbans region. Those within 1 km of the coast or river face more significant cyclonic impact, leading to more pronounced property loss and, consequently, a higher propensity to migrate. In contrast, households situated further inland, beyond

Table 5.10 Cyclonic impact

Predictor	Estimate	SE	Z	Odds ratio
1: Housing structure change within 1 km distance, 0: Other condition				
1–0	1.081	0.357	3.0247	0.339**

Note ** $p < 0.01$; *** $p < 0.001$ (*Source* Computed by the researcher based on primary survey)

1 km, are relatively insulated from the direct effects of cyclonic storms, reducing their migration likelihood.

This study highlights the multifaceted nature of migration decisions in the face of environmental hazards. The findings emphasize the importance of considering both the physical proximity to danger zones and the resultant property damage when assessing the drivers of migration. These insights are critical for policymakers aiming to devise targeted interventions to mitigate the adverse impacts of cyclones on vulnerable populations. The logistic regression results underscore the significant influence of cyclonic vulnerability on migration decisions in the coastal Sundarbans region. Households within 1 km of the coast or river, particularly those experiencing property losses, exhibit a markedly higher propensity to migrate. Addressing these vulnerabilities through improved infrastructure and targeted support could significantly reduce the need for migration and enhance the resilience of affected communities (Table 5.10).

5.5 Discussion

The movement of people from one place to another, known as human migration, has been significantly shaped by changes in climate and weather patterns. Scholars like McLeman (2013), McLeman et al. (2008), and McLeman and Smit (2006) have extensively explored this connection. Climate variability directly impacts various aspects of human life, including health, finances, and social activities, influencing migration trends (Dallmann and Millock 2017). In regions like the Sundarbans, where cyclones are frequent, migration patterns are particularly affected. Researchers have conducted empirical studies to understand this phenomenon better. They employed various data collection methods, such as structured questionnaires, personal interviews, and focus group discussions. Additionally, Information was gathered via phone from migrants who were absent during household surveys. To analyze the factors influencing migration, researchers used logistic regression. They classified households based on whether or not one or more individuals had migrated (considered as migrated households) (non-migrated households). Other relevant variables, identified through literature review and pilot surveys, were also considered. This research sheds light on the complex interplay between climate change, socio-economic factors, and migration patterns, providing valuable insights for understanding and addressing contemporary challenges in human mobility.

In severe events like droughts, out-migration intensifies significantly, often becoming a last resort for affected individuals. However, alongside these extreme situations, a regular pattern of seasonal and temporary male migration is observed annually, typically linked to the availability of work opportunities in destination areas. This phenomenon has been extensively studied by researchers such as Gupta et al. (2011), Keshri and Bhagat (2012). Interestingly, many of these migrants return home within six months, highlighting the transient nature of their movements. The decision to migrate is often driven by economic factors, with landless households seeking better living standards elsewhere. Indebtedness is a significant influencing factor for seasonal out-migration across different landholding sizes. However, economically stable individuals, typically engaged in fishing cultivation, may also be forced to migrate following cyclones, aiming to repay debts incurred from their agricultural losses. Studies by Debnath (2021) emphasize the role of landholding size as a primary economic determinant of male migration in rural areas. Social networks play a crucial role in the migration decision-making process, with friends being the primary source of destination information, intermediaries or contractors, and relatives. Economically and socially marginalized groups, such as Scheduled Castes (SC) and Scheduled Tribes (ST), exhibit higher migration rates compared to forward castes, often due to their higher vulnerability and landlessness (Fig. 5.13). The study also highlights that households with minimal crop losses due to their smaller landholding capacity are more likely to migrate, indicating a direct correlation between agricultural productivity and migration propensity. Moreover, middle-aged males (35–60 years) show a higher probability of out-migration than younger males (15–24 years). Both literate and illiterate individuals demonstrate an equal likelihood of migration, suggesting that educational attainment does not significantly influence migration decisions (Debnath and Nayak 2022). Furthermore, frequent cyclones and subsequent crop failures exacerbate unemployment and decrease agricultural activity, further prompting migration. Overall, the findings underscore the multifaceted nature of migration dynamics, influenced by economic, social, and environmental factors, and emphasize the role of rural distress (Figure) and lack of opportunities in driving male out-migration from rural areas.

Gray and Mueller (2012) and Chen et al. (2017) have observed a paradox: while impoverished individuals are more susceptible to environmental factors, they are less likely to migrate due to limited resources. Some studies (Penning-Rowsell et al. 2013) suggest that impoverished households are compelled to move as a "last resort," often the most vulnerable to flooding. However, other research emphasizes that migration requires resources, and those with more resources are more likely to relocate, leaving people experiencing poverty or resource-deprived individuals trapped in their current locations, labeled as a "trap population" (Black et al. 2011) (Fig. 5.14). Despite their hazardous conditions, many people remain committed to their places of residence, which deters migration (Bernzen et al. 2019). Quantifying the exact impact of migration is challenging. Black et al. (2011) theoretically established that the complex conditions influenced directly by the environment and indirectly by socio-economic factors could be affected by cyclonic hazards indirectly. While it may be possible to measure the direct impact of the environment on migration, quantifying the indirect

Fig. 5.13 A woman from the Sundarbans ventures into the river with her children to catch crabs, as her husband is physically unfit to work. In this climate-vulnerable region, women often take on the burden of both caregiving and livelihood, navigating dangerous waters to sustain their families amidst rising ecological and economic hardships. *Source* By authors on 19-04-2023

effects poses a challenge. A study (Paul et al. 2022) revealed that 10% of migrations occurred due to severe cyclones that destroyed farmland and housing. These migrations represent a highly vulnerable segment of the population, reporting significant disruptions in household income and loss of assets.

The demographic most inclined toward migration comprises economically disadvantaged individuals, particularly those without land ownership, such as daily laborers. Primarily, younger males seek opportunities to earn income to support their families through remittances (Hunter et al. 2015). This inclination toward migration is exacerbated by environmental degradation, such as destroying neighboring farmland due to salt leaching, which weakens physical resilience against storms and flooding (Azam 2011). Research confirms that coastal villages in the Sundarbans are particularly susceptible to migration due to frequent cyclonic activity, which damages housing structures and economic livelihoods. Living farther from rivers or coastlines decreases the likelihood of migration (0.529 for distances of 500 m–1 km compared to <500 m). Despite the significant impact of environmental hazards, the study suggests that economic and social factors play a more substantial role in driving migration decisions. Gray and Mueller (2012) concluded that while flooding may lead to short-term displacement, there is limited evidence of long-term migration resulting from such events.

Fig. 5.14 As he describes the loss of his family property, a coastal inhabitant conveys a sense of both sadness and danger. Unrelenting sea waves have degraded between 500 and 700 m of land over the last 20 years. Due to financial difficulties, his family still resides in the neighborhood despite the danger, underscoring their fragility and entrapment in a dangerous setting. *Source* By authors on 18-11-2023

Similarly, Yan et al. (2014) found that households with larger farms are less likely to migrate. Furthermore, the study highlights several demographic factors associated with migration likelihood. Nuclear households are 0.449 times less likely to migrate than joint families. Additionally, larger families (with 2–3 members) and those with 2–3 earning members exhibit 2.24- and 4.2-times higher odds of migration, respectively, compared to smaller families (with one member) and those with only one earning member. These findings underscore the complex interplay of economic, social, and environmental factors in shaping migration patterns and emphasize the importance of considering various demographic characteristics when examining migration trends.

As an adaptation strategy to climate change, migration is a significant and multifaceted phenomenon in India and Bangladesh. Both countries face severe climatic challenges that drive internal and cross-border migration. This discussion delves into the specific contexts and research findings from these regions to illustrate how migration is an adaptive response to environmental stressors. Sundarban, with its low-lying geography, is particularly susceptible to climate change impacts such as sea-level rise, cyclones, and riverbank erosion. Research by Ahamad and Alam (2023) shows that, migration is a common adaptive strategy in coastal areas affected by sea-level rise and increased salinity. Families often move to nearby urban centers such as Dhaka or Chittagong, seeking better economic opportunities and safer living conditions. Long-term displacement is a significant issue in the context of slow-onset events like riverbank erosion in the Brahmaputra and Meghna Rivers. Rahman and Hossain (2023) note that environmental degradation forces communities to migrate permanently to urban areas, where they face new challenges, such as inadequate housing and limited access to essential services. Despite these challenges, remittances from urban migrants play a crucial role in supporting the rural economy and enhancing the adaptive capacity of families remaining in vulnerable areas (Haque and Alam 2023). India experiences various climatic events, from severe droughts to intense floods, which substantially affect migration. Research indicates that environmental factors significantly influence migration patterns within India. For instance, a study by Singh and Das (2023) highlights those droughts in Maharashtra have led to increased rural-to-urban migration as farmers seek alternative livelihoods in cities. Additionally, floods in Assam and Bihar have forced many to relocate temporarily to safer areas, often resulting in long-term displacement due to the destruction of homes and agricultural lands (Das et al. 2023).

Social and economic factors also play crucial roles in migration decisions. In regions like West Bengal, where recurring floods affect agricultural productivity, families often send younger members to urban areas to diversify income sources and reduce dependency on vulnerable agricultural practices (Banerjee and Roy 2023). This form of migration helps families build financial resilience against future climatic shocks.

5.6 Conclusion

The research on the drivers behind out-migration from multi-hazard zones in the Indian coastal Sundarbans offers profound insights into the intricacies of human displacement driven by environmental and socio-economic factors. This study meticulously examined how migration patterns in the Sundarbans vary, particularly emphasizing the stark contrast between coastal and inland areas. The findings reveal that coastal and riverine zones experience significantly higher migration rates, primarily attributed to the increased exposure to cyclones in these regions. A critical observation from the study is the role of economic pressures in influencing migration decisions. Households with a more significant number of income-contributing members were found to be more likely to migrate. This tendency is driven by seeking better economic opportunities and improving living standards. In this context, migration serves as a vital coping mechanism, allowing families to transition from rudimentary living conditions in makeshift huts or mud houses to more resilient semi-cemented or fully cemented homes. The substantial losses in agricultural produce and fishing activities, exacerbated by cyclonic events, further catalyze the decision to migrate as families seek to mitigate their economic vulnerabilities.

The study employed logistic regression analysis to identify the key factors influencing migration. Geographic location emerged as a significant determinant, with households in coastal villages showing a higher propensity to migrate compared to those in inland regions. Family size and the number of wage earners within a household were crucial factors; larger families and those with multiple income sources were more inclined toward migration. The analysis also highlighted a notable social dimension, indicating that households belonging to Scheduled Castes (SC) exhibited greater odds of migrating than those from Unreserved Castes. This finding underscores the intersection of social stratification and vulnerability, suggesting that marginalized communities may face compounded pressures that drive migration. One of the pivotal insights from the study is the impact of housing conditions on migration decisions. Severe damage to homes due to cyclones was found to hinder mobility, illustrating a paradox where the very disasters prompting migration also constrain it. So, migration can be an effective strategy for coping with the aftermath of cyclones. The extent of damage incurred may limit a household's ability to relocate, trapping them in a cycle of vulnerability.

Overall, the research underscores the complex interplay of environmental, economic, and social variables that influence migration decisions in the Sundarbans. The decision to migrate is not solely a reaction to immediate environmental threats but is also deeply intertwined with broader socio-economic dynamics. This complexity highlights the necessity for targeted interventions aimed at supporting vulnerable households. Specialized aid and proactive measures are essential to protect these communities, ensuring they have the resources and support needed to relocate safely or enhance their resilience. The study's findings contribute to understanding migration in response to environmental risks. The research emphasizes the importance of holistic approaches in policy formulation by illuminating the nuanced dynamics at play. Addressing the multifaceted nature of migration requires coordinated efforts that integrate environmental management, economic support, and social protection measures (Fig. 5.15). This comprehensive approach can help mitigate the adverse impacts on affected communities, fostering resilience and sustainable development in regions like the Sundarbans that are susceptible to multiple hazards.

In conclusion, the study provides a detailed analysis of the drivers behind out-migration from the Indian coastal Sundarbans, presenting a sophisticated understanding of how various factors converge to influence this phenomenon. The research highlights the critical need for tailored interventions to address the unique challenges faced by these communities, ensuring that migration remains a viable and beneficial strategy for those seeking to escape the perils of multi-hazard zones. Through its in-depth exploration of migration dynamics, the study contributes significantly to the discourse on environmental migration, offering insights crucial for developing effective policies and programs to mitigate the impacts of environmental disasters on vulnerable populations.

Fig. 5.15 Left side stand the women with her younger son, whose husband went outside as a migrant worker but didn't come back home, and he also married a new wife in the migration destination, and the right side is the author (Pintu Mandal) with women' elder daughter who is also physically challengeable. *Source* By authors on 10-01-2024

References

Afifi T, Milan A, Etzold B, Schraven B, Rademacher-Schulzb C, Sakdapolrak P, Warner K (2016) Human mobility in response to rainfall variability: opportunities for migration as a successful adaptation strategy in eight case studies. Migrat Develop 5(2):254–274

Ahamad T, Alam S (2023) Coastal migration in Bangladesh: adapting to sea-level rise and salinity intrusion. Bangl Clim Res J 9(3):101–118

Azam M (2011) Factors driving environmentally induced migration in the coastal regions of Bangladesh: an exploration study. Verlag Nichtermittelbar

Banerjee S, Roy A (2023) Floods and migration in West Bengal: socio-economic impacts and adaptation strategies. Indian J Environ Stud 18(1):33–50

Bernzen A, Jenkins JC, Braun B (2019) Climate change-induced migration in coastal Bangladesh? A critical assessment of migration drivers in rural households under economic and environmental stress. Geosciences 9(1):51

Bhagat RB, Keshri K, Ansary R (2018) Internal migration in India: Intensity, flows and impact. In: Comparing internal migration in the countries of Asia' conference. Asian Demographic Research Institute, Shanghai University, Shanghai, China

Biswas B, Mallick B (2021) Livelihood diversification as key to long-term non-migration: evidence from coastal Bangladesh. Environ Dev Sustain 23(6):8924–8948

Black R, Bennett SR, Thomas SM, Beddington JR (2011) Migration as adaptation. Nature 478(7370):447–449

Chen JJ, Mueller V, Jia Y, Tseng SKH (2017) Validating migration responses to flooding using satellite and vital registration data. Am Econ Rev 107(5):441–445

Dallmann I, Millock K (2017) Climate variability and inter-state migration in India. Cesifo Econ Stud 63(4):560–594

Das S, Islam ARMT, Kamruzzaman M (2023) Assessment of climate change impact on temperature extremes in a tropical region with the climate projections from CMIP6 model. Clim Dyn 60(1):603–622

Debnath M (2021) Exploring the behaviour of seasonal agricultural migrants in the Rarh region of West Bengal, India. Spat Inf Res 29(1):83–96

Debnath M, Nayak DK (2022) Assessing drought-induced temporary migration as an adaptation strategy: evidence from rural India. Migr Dev 11(3):521–542

Daoud M (2016) Living on the edge: Gender relations, climate change and livelihoods in the villages of Maryut and Nubia, Egypt (Doctoral dissertation, University of East Anglia)

Gray CL, Mueller V (2012) Natural disasters and population mobility in Bangladesh. Proc Natl Acad Sci 109(16):6000–6005

Gupta AK, Tyagi P, Sehgal VK (2011) Drought disaster challenges and mitigation in India: strategic appraisal. Curr Sci 1795–1806

Haque M, Alam M (2023) The role of remittances in rural adaptation to climate change in Bangladesh. Int J Migr Dev 15(2):85–100

Hunter LM, Luna JK, Norton RM (2015) Environmental dimensions of migration. Ann Rev Sociol 41:377–397

Hollowed AB, Haynie AC, Hermann AJ, Holsman KK, Punt AE, Szuwalski CS (2022) Implications of climate change on the Bering Sea and other cold water systems. Deep Sea Research Part II: Topical Studies in Oceanography 201:105110

https://www.internal-displacement.org/global-report/grid2024/

IDMC (2021) https://www.internal-displacement.org/global-report/grid2021/

Keshri K, Bhagat RB (2012) Temporary and seasonal migration: Regional pattern, characteristics and associated factors. Econ Polit Weekly 81–88

Keshri K, Bhagat RB (2013) Socio-economic determinants of temporary labor migration in India: a regional analysis. Asian Popul Stud 9(2):175–195

McAuliffe M, Ruhs M (2017) World migration report 2018. Geneva: International Organization for Migration, pp 1–32

McLeman R (2013) Developments in modelling of climate change-related migration. Clim Chang 117:599–611

McLeman R (2018) Thresholds in climate migration. Popul Environ 39(4):319–338

McLeman R, Smit B (2006) Migration as an adaptation to climate change. Clim Chang 76(1):31–53

McLeman R, Mayo D, Strebeck E, Smit B (2008) Drought adaptation in rural eastern Oklahoma in the 1930s: lessons for climate change adaptation research. Mitig Adapt Strat Glob Change 13:379–400

Mondal M, Biswas A, Haldar S, Mandal S, Bhattacharya S, Paul S (2022) Spatio-temporal behaviours of tropical cyclones over the bay of Bengal Basin in last five decades. Trop Cycl Res Rev 11(1):1–15

Naser MM, Swapan MSH, Ahsan R, Afroz T, Ahmed S (2019) Climate change, migration and human rights in Bangladesh: perspectives on governance. Asia Pac Viewp 60(2):175–190

Parganiha OP, Sharma ML (2017) Farmer's perception about climate change in plain zone of Chhattisgarh. Indian Res J Extens Educ 17(3):92–96

Paul BK, Rahman MK, Lu M, Crawford TW (2022) Household migration and intentions for future migration in the climate change vulnerable lower Meghna estuary of coastal Bangladesh. Sustainability 14(8):4686

Penning-Rowsell EC, Sultana P, Thompson PM (2013) The 'last resort'? Population movement in response to climate-related hazards in Bangladesh. Environ Sci Policy 27:S44–S59

Priodarshini R, Mallick B (2021) Challenges and potentials of (non-) migration AS climate change adaptation IN Bangladesh and beyond. Plan Plus 11(1)

Rahman M, Hossain Z (2023) Riverbank erosion and long-term displacement: case studies from the Brahmaputra and Meghna Rivers. Environ Displac J 7(2):73–88

Sengupta A, Samanta G (2022) Understanding immobility of a highly vulnerable coastal village in the Indian Sundarban. Reg Environ Change 22(3):90

Singh R, Das S (2023) Environmental migration in Maharashtra: an analysis of rural-urban dynamics. J Clim Change Dev 12(2):45–62

Twinomuhangi R, Sseviiri H, Kato AM (2023) Contextualising environmental and climate change migration in Uganda. Local Environ 28(5):580–601

Viswanathan B, Kumar KK (2015) Weather, agriculture and rural migration: evidence from state and district level migration in India. Environ Dev Econ 20(4):469–492

Yan X, Bauer S, Huo X (2014) Farm size, land reallocation, and labor migration in rural China. Popul Space Place 20(4):303–315

Chapter 6
Major Findings, Policy Recommendation, and Conclusion

Abstract Between 1990 and 2023, research on climate migration grew significantly, especially after the IPCC Fourth (2007) and Fifth (2014) Assessment Reports. Keyword analysis highlights major themes like migration, climate change, and adaptive management. Future research must adopt interdisciplinary, holistic approaches to address the complex economic, health, and environmental impacts of climate displacement. Findings show that migration is often a response to short-term threats rather than long-term adaptation. Cyclones have severely impacted fishing, aquaculture, shrimp farming, and agriculture, mainly due to prolonged waterlogging and increased soil salinity, leading to crop failures and income loss. Many affected households resorted to borrowing, though migrant households repaid debts faster than non-migrants. Migration rates were higher in coastal villages, among Scheduled Castes (SC), larger families, and households with multiple earners. However, severe housing damage often trapped the poorest in hazardous areas, limiting their ability to relocate. Migration should be recognized as a planned climate adaptation strategy, ensuring it remains safe, voluntary, and dignified. Policies must promote livelihood diversification, strengthen receiving areas, protect cultural identity through community relocation, and address the needs of trapped populations with in situ adaptation, building resilience among vulnerable coastal communities facing cyclonic hazards.

Keywords IPCC · Trapped population · Climate-induced migration · Voluntary migration · Livelihood diversification · Policy recommendation · Hierarchy

6.1 Introduction

Climate migration has emerged as an urgent issue confronting our global society, driven by the intensifying impacts of climate change, including rising sea levels, severe weather events, and prolonged droughts. This phenomenon, termed climate-induced displacement, refers to the forced migration of individuals and communities who must abandon their homes due to the adverse effects of climate change on their living conditions. The consequences of climate migration are profound, affecting

P. Mandal and S. Paul, *Climatic Vulnerability and Migration: Evidence from Indian Sundarbans*, Springer Geography, https://doi.org/10.1007/978-3-032-05858-4_6

not only the migrants themselves but also the regions they leave and the destinations they move to. This study aims to delve into the complexities of climate migration, evaluate current strategies, and propose recommendations for effectively addressing this multifaceted problem.

The Intergovernmental Panel on Climate Change (IPCC) has underscored the urgency of addressing migration induced by climate change. They project that by 2050, tens of millions of people may be compelled to relocate due to climate-related challenges (Arias et al. 2021). This crisis predominantly impacts coastal regions, particularly those at low elevations. A notable example of an ecologically vulnerable area facing substantial risks is the Indian Sundarbans. This region frequently experiences cyclones, resulting in large-scale community displacements (Bhattacharyya and Werz 2012). The Cyclone-Induced Migration Management Policy for the Sundarbans provides a comprehensive framework that includes measures to prevent displacement, protect individuals during displacement, and establish long-term solutions (Gakpey 2022).

The primary objective of climate migration policies is to avert relocation by implementing proactive measures that mitigate vulnerability to climatic disasters. Essential strategies include conducting risk assessments, developing early warning systems, and integrating climate risk considerations into development planning to reduce the impact of climate change on vulnerable communities (de Siqueira et al. 2022). Additionally, it is imperative to allocate resources toward the development of resilient infrastructure and the promotion of sustainable land use practices as critical components of these preventive measures (World Bank 2013).

By focusing on these aspects, we can create a more resilient global society capable of withstanding the challenges posed by climate-induced displacement. This requires not only a comprehensive understanding of the dynamics at play but also a concerted effort to implement policies and practices that safeguard both individuals and communities from the escalating threats of climate change. Through collaboration and innovative solutions, we can address the pressing issue of climate migration and ensure a sustainable future for all.

Ensuring the safety, dignity, and rights of displaced individuals is paramount during periods of displacement. Rapid humanitarian interventions, effective partnerships with non-governmental organizations, and the protection of fundamental human rights are essential to delivering critical assistance to displaced communities (Macintyre 2020). A key strategy outlined in existing regulations involves the establishment of temporary relief camps that provide essential services and legal assistance to help individuals reclaim their property and access government programs (Mashi et al. 2019).

Long-term strategies for displaced populations include facilitating secure and voluntary repatriation, providing support for local integration, and relocating groups to safer areas. These measures aim to create sustainable living conditions and employment opportunities for those who have been displaced, ensuring their well-being and fostering social cohesion with host communities (IOM 2018). Policies must incorporate robust institutional frameworks and reliable financial mechanisms to effectively implement these measures and support the affected communities (UNHCR 2021).

The study of climate migration from 1990 to 2023 has seen significant advancements, characterized by notable milestones and increased scholarly interest. Initially, publications and data on the topic were sparse. However, the publication of the IPCC Fourth Assessment Report in 2007 and the IPCC Fifth Assessment Report in 2014 significantly boosted the volume of publications and data availability. Post-2014, there was a steady rise in research papers, reflecting a heightened global awareness and stronger support for research initiatives. Advancements in data collection and processing tools enabled researchers to conduct more comprehensive empirical studies and employ sophisticated quantitative methods. The focus shifted from theoretical frameworks to evidence-based research, underscoring the need for rigorous empirical investigations to inform policy and practice. By emphasizing these strategies and advancements, we can better address the complex challenges posed by climate-induced displacement. It is crucial to adopt a holistic approach that not only addresses immediate humanitarian needs but also promotes long-term resilience and integration. Through collaborative efforts and innovative solutions, we can ensure the protection and empowerment of displaced individuals, fostering a more inclusive and resilient global community capable of withstanding the impacts of climate change.

The VOSviewer analysis underscores the preeminent role of the United States in climate migration research, with substantial contributions also emanating from China, the United Kingdom, and France. In contrast, peripheral nations exhibit lower levels of research productivity, highlighting disparities in research capabilities and the distribution of funding. Keyword analysis reveals significant focal points such as climate change, migration, and adaptive management, indicating primary areas of scholarly attention. Future research objectives and imperatives involve fostering interdisciplinary collaboration and adopting holistic approaches to address the economic and health-related consequences of climate migration, as well as the complex challenges posed by environmental displacement. Establishing a robust nexus between research, policy, and the human dimensions of climate change is critical for effective governance and enhancing community resilience.

Identifying nodes with fewer connections or smaller clusters in the co-citation network is crucial for pinpointing prospective research gaps. For instance, if specific economic or health-related terms show minimal association, it suggests that these areas are underexplored and warrant further investigation. By addressing these deficiencies, scholars can concentrate their efforts on these neglected topics, potentially uncovering new insights and making significant contributions to the field. A thorough understanding of the core concepts and their interrelationships within the network can provide valuable guidance for charting future research directions. Strengthening the connection between policy and human dimensions, for example, could yield valuable knowledge on effective governance strategies and bolster community resilience.

Researchers can develop more comprehensive strategies for managing the socio-economic and environmental challenges posed by climate change by examining how policies influence human responses to climate migration. Embracing an interdisciplinary approach will facilitate the creation of innovative solutions and enhance our holistic understanding of climate migration. This, in turn, will enable the formulation of more robust and sustainable strategies to address this global issue. By integrating diverse perspectives and methodologies, we can better tackle the multifaceted problems associated with climate-induced displacement, ensuring that communities worldwide are better equipped to adapt and thrive amid the evolving challenges of climate change.

6.2 Summary of Major Findings from the Research Objectives

The study tries to explain that the study used well-established and reliable methods and procedures to conduct its analysis. These traditional techniques are chosen specifically to ensure the results are credible and dependable. The important findings from this analysis will be discussed in the next section, which focuses on the objectives of the research.

6.2.1 Objective 1 (To Mapping the Cyclone-Induced Potential Human Migration Risk in Pathar Pratima CD Block, South 24 Parganas District, West Bengal)

Throughout history, migrations have significantly shaped the world, often driven by climate threats and environmental changes. The International Organization for Migration (IOM) defines environmental migration as the movement of people caused by sudden or gradual environmental changes that negatively affect their quality of life. This movement can be voluntary or involuntary, domestic or international, and temporary or permanent (IOM 2011). Various studies, including those by IOM (2018), Jha et al. (2018), and Luetz and Merson (2020), have explored migration as both a consequence of climate change and an adaptation strategy. Conversely, other scholars argue that environmental and climate changes act as stressors that exacerbate migration, as suggested by Scheffran et al. (2012) and Kenfack (2022).

The study highlights that cyclonic risks are the most significant drivers of human mobility in the examined region. In this context, migration decisions are predominantly a coping mechanism rather than a proactive adaptation strategy. This suggests that individuals move primarily to escape immediate threats rather than to adapt to long-term changes. Consequently, the study underscores numerous unresolved issues surrounding climate-induced migration. It offers guidance to academics, governmental organizations, and non-governmental groups on selecting survey areas for empirical studies, aiming to address the complexities of climate-induced migration effectively.

One major finding is the dual role of migration as both a response to and a driver of climate change. This nuanced perspective is supported by a body of research, indicating that while migration can be an adaptation strategy, it can also contribute to the challenges of climate change. This complexity necessitates a deeper understanding of the interplay between migration and environmental factors.

Another significant finding is the emphasis on environmental and climate changes as amplifiers of migration pressures. Scholars argue that these changes create stressors that force people to move, often exacerbating existing vulnerabilities. This perspective aligns with the view that environmental factors alone are rarely the sole cause of migration but interact with socio-economic and political factors to drive human mobility.

The study also identifies cyclonic threats as the primary catalyst for migration in the studied area. This finding is crucial for policymakers and practitioners as it highlights the immediate need to address cyclonic hazards to mitigate forced migration. Enhancing resilience to cyclones through improved infrastructure, early warning systems, and community preparedness can reduce the necessity for migration as a coping mechanism.

6.2.2 *Objective 2 (To Comprehend the Nature of Migration and the Demographic as Well as Socio-economic Condition of Coastal Dwellers)*

This study investigates the socio-economic repercussions of the 2020 Cyclone Amphan on households in the southern coastal areas of West Bengal, focusing on both migratory and non-migratory communities. Conducted in the Pathar Pratima CD Block of the South 24 Parganas District, the study uncovers the substantial impacts of Amphan on these coastal villages, driving some residents to seek alternative housing solutions. The findings starkly illustrate how the cyclone disrupted livelihoods, income sources, housing stability, food accessibility, and agricultural productivity.

One major finding is the cyclone's devastating effect on agriculture, fishing, aquaculture, and shrimp farming, primarily due to prolonged water logging, which elevated saline levels in soil and water bodies. These adverse conditions led to crop failures and diminished yields, compelling households to find alternative income sources. The timing of the cyclone, just before the harvest, inflicted severe financial losses on farmers, many of whom were forced to relocate in search of support. This highlights the critical vulnerability of agricultural livelihoods to extreme weather events and the ensuing economic displacement.

Another significant finding is the shift in land ownership dynamics among migrant households. There was a notable decrease in small landowners and a corresponding increase in landless and marginal farmers. The cyclone's breach of protective embankments exposed agricultural lands and shrimp farms to flooding and heightened salinity, exacerbating the situation. The protracted process of rebuilding these embankments prolonged soil salinity issues, rendering recovery efforts futile for certain farms even 2 years post-Amphan. This underscores the long-term challenges of post-disaster recovery and the need for more resilient infrastructure.

The study also reveals that affected households resorted to borrowing from both official and unofficial sources to cope with the financial strain. Interestingly, migrant households exhibited a significantly higher rate of debt repayment compared to non-migrant households. This suggests that migration may be a viable strategy for improving financial resilience, as it offers opportunities to increase income or alleviate financial burdens, such as loan repayments. Additionally, there was a marked decrease in the number of working days per month for migrant households' post-Amphan, indicating that migration can also serve as a practical approach to stabilize income during crises.

Food insecurity emerged as a critical issue following the cyclone, with a significant increase in famine conditions. However, migrant households demonstrated higher levels of food consumption compared to non-migrant households, indicating that migration and occupational changes had a positive impact on dietary habits. This finding suggests that migration can enhance food security and overall well-being in the aftermath of climate-related disasters.

In conclusion, the study posits that migration or livelihood diversification can be effective strategies for coping with climate-related disasters such as Cyclone Amphan. Migrant households were found to be more resilient in terms of income, housing, food security, and debt repayment. These findings underscore the importance of developing adaptive strategies that include migration as a viable option for enhancing the resilience of vulnerable communities to future climatic events.

6.2.3 Objective 3 (To Review the Determinants of Migrants, with a Specific Focus on the Impacts of Cyclonic Events)

This study aimed to unravel the factors prompting migration from multi-hazard zones in the Indian coastal Sundarbans, revealing critical insights into the dynamics influencing this movement. The data indicated stark variations in migration rates based on proximity to the coast, with coastal and riverine regions exhibiting higher migration due to heightened vulnerability to cyclones. A notable finding is that households with a greater number of income earners were more inclined to migrate, driven predominantly by economic constraints. Migration emerged as an adaptation strategy, enabling households to upgrade their living conditions from fragile huts or mud houses to more durable semi-cemented or fully cemented dwellings. This shift underscores the role of migration in enhancing household resilience against environmental threats.

Furthermore, the study highlighted the substantial impact of cyclonic damages on agricultural and fishing activities as a crucial determinant of migration decisions. Logistic regression analysis was employed to identify key factors influencing migration, including geographic location, family size, and number of wage earners, housing conditions, and cyclone-induced damages. Coastal villages displayed a stronger propensity for migration compared to inland areas, while households belonging to Scheduled Castes (SC) showed a higher likelihood of movement than those from Unreserved Castes. Additionally, larger families and those with multiple wage earners were more prone to migrate, emphasizing the interplay between economic capacity and migration decisions.

A significant revelation of the study is the impediment posed by substantial house damage on mobility. While migration can aid households in coping with the effects of cyclones, severe damage to homes restricts their ability to relocate, trapping them in hazardous conditions. This finding underscores the paradox where those most in need of moving away from danger zones are often the least capable of doing so due to the loss of shelter.

In summary, the study elucidates the complex amalgamation of environmental, economic, and social factors influencing migration in the Sundarbans. The combination of heightened cyclone vulnerability, economic necessity, and social dynamics drives the decision to migrate, with proximity to the coast and household composition playing pivotal roles. The findings emphasize the critical need for targeted assistance to protect at-risk households in these perilous zones. By providing a nuanced understanding of the migration dynamics in response to environmental threats, the study underscores the importance of preemptive measures to mitigate the adverse effects on impacted communities. Ultimately, the study offers unique insights into the multifaceted nature of migration as an adaptive response to environmental hazards, highlighting the need for comprehensive strategies that address the root causes and barriers to mobility. It calls for a balanced approach that combines immediate relief with long-term resilience-building initiatives to support the most vulnerable populations in the Sundarbans.

6.3 Policy Recommendation

Policy planning for cyclone-induced migration in coastal India is crucial for several reasons. Coastal regions are increasingly vulnerable to extreme weather events due to climate change, and cyclones often trigger significant population displacement. Effective policy planning helps mitigate the adverse effects of such migration by ensuring timely and coordinated responses. It enables the development of infrastructure and systems that support displaced populations, including emergency shelters, health services, and livelihood opportunities. Furthermore, well-crafted policies facilitate risk assessment and management, helping communities adapt to changing conditions and reducing the need for future migration. By addressing the socio-economic impacts of displacement, such as loss of income and disruption of social networks, policy planning supports resilience-building efforts and ensures that affected populations can recover more swiftly. Thus, strategic policy intervention is essential for safeguarding the well-being of coastal communities and enhancing their ability to withstand and adapt to cyclone-induced challenges. In this research, some policy planning approach has been made based on the outcomes of specific objectives within the study region.

The study specifically highlights the Pathar Pratima block, especially its southern and eastern borders, as the most susceptible to these risks, with embankment ruptures being the primary determinant.

6.3.1 Specific Policy in Perspective to Cyclonic Exposure

Based on these findings, the following specific policy planning measures are recommended:

Level 1 at Primary Hierarchy

Mangrove Plantation: Initiate comprehensive mangrove afforestation initiatives to mitigate coastal erosion and establish natural barriers against cyclonic storms. **Livelihood Diversification**: Initiate government schemes to provide alternative livelihoods, such as aquaculture, handicrafts, and ecotourism, to reduce dependency on agriculture and enhance economic resilience.

Level 2 at Secondary Hierarchy

Embankment Reinforcement: Prioritize the construction and maintenance of high-quality embankments using durable materials. Implement regular inspections and prompt repairs to prevent breaches during cyclonic events. **Community Awareness and Training**: Conduct community-based disaster risk reduction programs to educate residents about emergency preparedness, early warning systems, and sustainable agricultural practices.

Level 3 at Tertiary Hierarchy

Infrastructure Development: Develop resilient infrastructure, including the construction of cyclone-resistant houses and an improved road network to ensure accessibility and reduce isolation during disasters. **Integrated Risk Management**: Develop an integrated risk management framework that combines environmental conservation, infrastructure development, and socio-economic strategies to create a holistic approach to disaster resilience. These policy measures are illustrated in the following diagram, which highlights the interconnected strategies required to enhance the resilience of coastal communities in Fig. 6.1. By implementing these policies, the region can significantly reduce its vulnerability to multiple hazards, ensuring sustainable development and improved quality of life for its inhabitants.

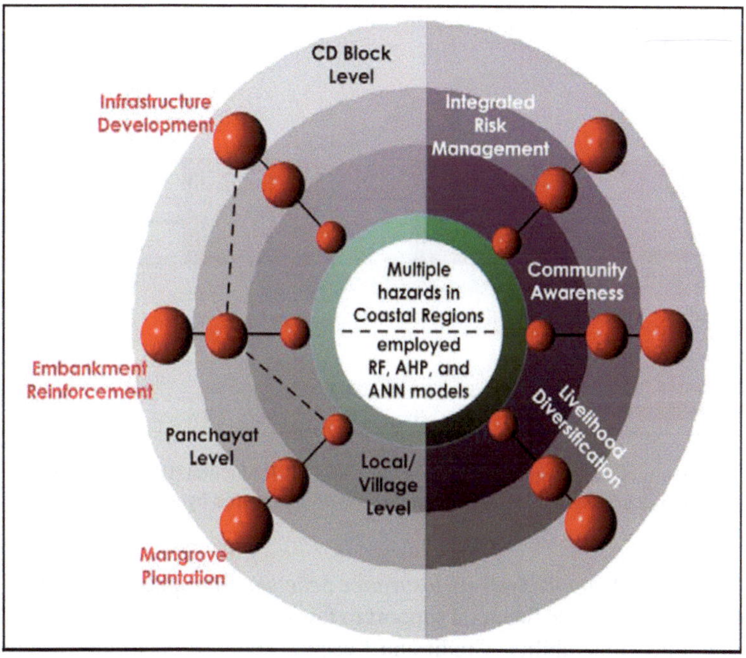

Fig. 6.1 Core components for addressing climate-induced cyclonic hazard at different hierarchies with integrated risk management (*source* prepared by authors)

6.3.2 Specific Policy in Perspective to Livelihood Vulnerability

The coastal Sundarbans region faces significant agricultural challenges due to cyclonic threats, disrupting paddy crop cultivation. These cyclones cause soil infertility, economic instability, and forced migration. The region also faces increasing concerns about post-monsoon water scarcity and soil salinity, particularly in southern mouzas of the Pathar Pratima CD Block, which affect crop diversity and cultivation. To address these challenges, a multi-tiered policy planning approach is recommended, illustrated in the flow model (Fig. 6.2).

Disaster Risk Reduction (DRR) Early Warning Systems: Develop and deploy advanced cyclone forecasting systems to provide timely warnings and facilitate preemptive measures. **Infrastructure Resilience**: Invest in cyclone-resistant infrastructure, including elevated farm buildings and improved drainage systems to minimize flood impact. *Agricultural Adaptation Strategies Salinity Management:* Implement soil salinity mitigation techniques such as saline-resistant crop varieties, improved irrigation practices, and soil conditioning to manage and reduce salinity levels. *Diversification:* Promote crop diversification and the adoption of climate-resilient farming practices to reduce dependency on a single crop and enhance soil

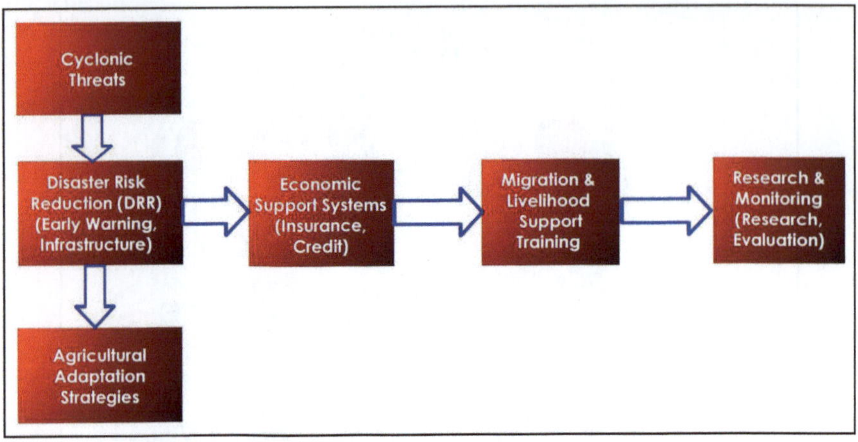

Fig. 6.2 Agricultural adaptation strategies for Disaster Risk Reduction (DRR) acquiring alternative livelihoods and integrating into new communities (*source* prepared by authors)

health. *Economic Support Systems Insurance Schemes*: Introduce agricultural insurance programs to provide financial support to farmers affected by cyclones, reducing their economic vulnerability. Credit and Leasing Support: Facilitate easier access to credit and favorable leasing agreements to help farmers recover and sustain their farming activities post-disaster. *Migration and Livelihood Support Skill Development*: To assist migrants in coastal areas, provide programs for skill development and vocational training. *Community-Based Support*: Establish community support networks to assist displaced individuals and provide resources for their resettlement and adaptation. *Research and Monitoring Ongoing Research*: Support research on climate impacts, soil health, and crop resilience to continuously adapt and refine agricultural practices. *Monitoring and Evaluation*: Regularly assess the effectiveness of implemented policies and adapt strategies based on real-time data and feedback.

Addressing the agricultural fragility in the Sundarbans region requires a comprehensive approach that integrates disaster risk reduction, adaptation strategies, economic support, migration assistance, and continuous research. By implementing these policies, it is possible to enhance agricultural resilience and support the livelihoods of coastal communities in the face of escalating climatic threats.

6.3.3 Specific Policy Concern with the Climate-Induced Migration

The Indian coastal Sundarbans experience a complex interplay of environmental, economic, and social factors that influence migration. Coastal and riverine zones are more vulnerable to cyclones, leading to higher migration rates. Households with multiple income earners are more likely to migrate due to economic constraints.

Cyclone-induced impacts, such as damage to agricultural and fishing activities, also influence migration decisions. Key migration determinants include geographic location, family size, number of wage earners, housing conditions, and cyclone-induced damages. Policies should address these barriers to enhance resilience and adaptive capacity in high-risk areas.

To address these issues effectively, a comprehensive policy planning framework is proposed. This framework integrates preemptive measures, immediate relief, and long-term resilience-building strategies (Fig. 6.3) which are described below.

The graphic presents policy recommendations aimed at addressing the vulnerabilities and migration risks faced by villages in cyclone-prone areas of West Bengal, particularly along the coastlines. The plan is divided into short-term and long-term strategies, aiming to mitigate immediate risks while ensuring sustainable solutions for the future. Short-term policy recommendations include disaster risk reduction, livelihood assistance, temporary shelter solutions, sustainable infrastructure investment, skill development and livelihood diversification, and resettlement programs. Disaster risk reduction involves strengthening early warning systems and enhancing embankment infrastructure to minimize the risks posed by cyclones and floods. Livelihood assistance includes credit access and asset replacement programs, particularly for family's dependent on agriculture and fishing. Temporary employment schemes should focus on public works related to disaster management, offering financial security and helping rebuild community resilience post-disaster. Long-term strategy recommendations include sustainable infrastructure investments, skill development and livelihood diversification, and resettlement programs for communities in vulnerable zones.

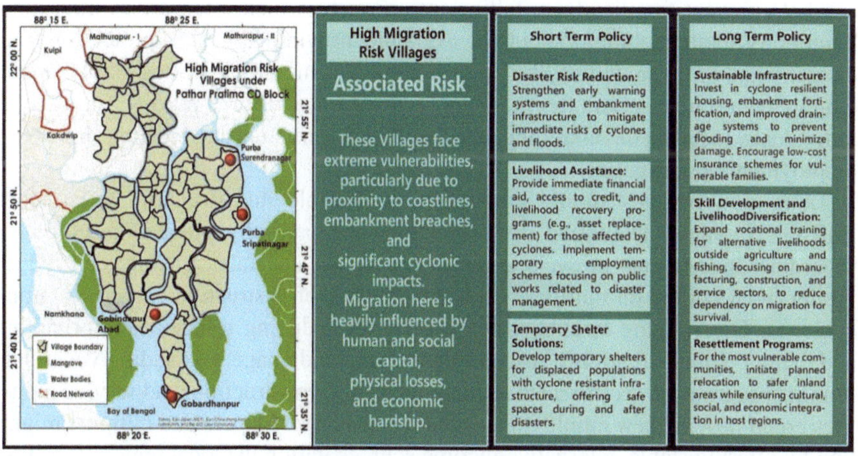

Fig. 6.3 Policy planning framework for sustainable development of the migrants under high migration risk villages under of Pathar Pratima CD Block, South 24 Parganas District, India (*source* prepared by authors)

Investment in cyclone-resistant housing and infrastructure, as well as low-cost insurance schemes, can help reduce long-term damage from cyclones and provide a financial safety net for vulnerable families. Vocational training programs should be expanded to encourage alternative livelihoods outside traditional sectors, such as manufacturing, construction, and service industries. Resettlement programs should be carefully planned to ensure community cohesion and avoid social dislocation. The policy framework addresses the high migration risks associated with identified villages due to their proximity to coastlines, frequent embankment breaches, and significant exposure to cyclonic impacts. By focusing on both structural and social interventions, these policies aim to reduce future risks and offer sustainable solutions for the affected populations.

Policy planning for moderate and low migration risk villages in cyclone-prone areas of West Bengal requires a comprehensive, phased approach that combines immediate and long-term strategies. The primary goal is to reduce the vulnerability of these villages to cyclones, minimize migration pressures, and ensure long-term resilience. The policy framework outlined can guide this process, focusing on disaster risk reduction, livelihood diversification, and infrastructure development while considering both short- and long-term needs.

In the short term, policy efforts should prioritize disaster risk reduction, livelihood assistance, temporary shelter solutions, and public works programs that offer immediate relief to affected populations. Enhancing early warning systems and improving embankment infrastructure is crucial for minimizing immediate risks posed by cyclones and floods. Livelihood assistance is essential for villages dependent on agriculture and fishing, which are highly vulnerable to cyclone damage. Policies that ensure quick access to credit and support asset replacement programs can help prevent families from falling into poverty, which often drives migration. Temporary shelter solutions should be available, providing physical protection from disasters and basic amenities like clean water and sanitation facilities. Public works employment schemes can offer financial security and a sense of purpose for affected individuals, focusing on rebuilding community infrastructure and creating short-term employment opportunities.

Long-term policies should emphasize sustainable infrastructure investments, skill development, livelihood diversification, and carefully planned resettlement programs. Investments in cyclone-resistant housing and infrastructure can reduce the financial burden on families and lower the risk of destruction caused by future cyclones. Diversifying livelihoods is crucial for reducing reliance on agriculture and fishing, which are highly susceptible to cyclone damage. Expanding vocational training programs in sectors such as manufacturing, construction, and service industries could provide new employment opportunities for villagers and mitigate migration by offering stable, alternative income sources within the local region. Resettlement programs may be necessary for the most vulnerable villages located in high-risk zones, but they need to be implemented with careful consideration of community dynamics to maintain community cohesion and avoid social dislocation.

A well-rounded policy approach that integrates both short-term and long-term strategies is essential for addressing the migration risks faced by moderate and low migration risk villages in West Bengal's cyclone-prone areas. By focusing on disaster risk reduction, livelihood support, and infrastructure development, these policies can offer immediate relief and build long-term resilience, minimizing the need for migration while enhancing the well-being of affected populations. The ultimate objective is to ensure that these communities can thrive in their existing locations, despite the challenges posed by climate-related disasters.

6.3.4 Policy Concerns When Climate-Induced Migration Becomes an Adaptation Strategy

Migration is turning into an inevitable adaptation strategy in areas like the Sundarbans, where cyclonic dangers are getting worse as a result of climate change. However, displaced communities frequently face extreme hardships as a result of unplanned, spontaneous migration. Therefore, in order to turn climate-induced migration into a planned, respectable, and sustainable type of adaptation, a number of crucial policy issues need to be resolved.

Recognizing Migration as a Planned Adaptation Strategy: Instead, then being interpreted as an indication of a failure to adapt, policy frameworks must recognize that migration can be a valid and proactive adaptation response. Migration should be included as a component of climate resilience in national and subnational climate adaptation strategies, such as India's State Action strategies on Climate Change (SAPCCs). Instead of being the final option after a tragedy occurs, policies must guarantee that migration decisions are voluntary and founded on well-informed choices. **Ensuring Safe, Dignified, and Voluntary Migration**: Making sure that migration happens in a dignified and safe manner is one of the top priorities. This necessitates funding-aided evacuation initiatives, early warning systems, and relocation plans in advance. In order to prevent coastal residents from being forced into hazardous living circumstances in urban slums or informal settlements, the emphasis should be on encouraging voluntary migration as opposed to reactive displacement. **Promoting Livelihood Diversification**: One of the main causes of migration from the Sundarbans is the decline of traditional means of subsistence like farming and fishing. Policies must support livelihoods that are climate resilient in order to increase local adaptive capacity. This entails supporting ecotourism cantered around mangroves, saline-resistant agriculture cultivation, and alternative skill-building initiatives. Migration can be made a choice rather than a necessity by bolstering local economies. **Protecting Cultural and Social Identity**: Strong social networks and cultural identities are frequently upended by migration. In order to maintain social cohesiveness, policy interventions should support cluster relocation, in which communities relocate together to new communities. Programs for cultural preservation at relocation sites can also support displaced groups in feeling like they belong. **Addressing Trapped Populations**: It is critical to acknowledge that not everyone is capable of migrating. The elderly, women, children, and the very poor are among the most vulnerable groups that may find themselves "trapped" in high-risk areas. For these populations, policies must guarantee in situ adaptation measures, such as bolstering embankments, building cyclone-resistant housing, and enhancing access to necessary services.

6.4 Contribution to Research

The contributions of this research study are in manifolds. These are discussed below.

6.4.1 Contribution to Knowledge

The study "Cyclone-Induced Vulnerability and Migration in Indian Sundarban" contributes to various knowledge domains, including environmental and climate science, disaster management and resilience studies, migration and demography, socio-economic studies, public health, policy and governance, geospatial and technological applications, and interdisciplinary research. It enhances understanding of cyclone dynamics, ecosystem impacts, disaster management and resilience, migration and demographics, economic impact, public health consequences, and health infrastructure resilience. The study also provides insights into the impact of climate change on cyclone dynamics in the Bay of Bengal region, focusing on the mangrove ecosystem and its alterations in species composition, health, and biodiversity. It also contributes to disaster management and resilience studies, assessing the vulnerability of coastal communities to cyclones and offering data and case studies on effective community-based resilience strategies. The study also explores the health consequences of cyclones, including physical injuries and mental health issues, and the resilience of health infrastructure in cyclone-prone regions. It also evaluates existing policies and governance structures, providing evidence-based recommendations for improving disaster management policies and coastal zone management.

6.4.2 Contribution to Methodology

The study "Cyclone-Induced Vulnerability and Migration in Indian Sundarban" is comprehensive research that combines various methods to assess the complex impacts of cyclones. Key contributions include the development of comprehensive vulnerability indices and resilience metrics, a mixed method approach that combines quantitative and qualitative research methods, and the use of Geographic Information Systems (GIS) mapping and remote sensing to monitor environmental changes. Predictive modeling and simulation are also employed, with climate models and migration simulation models being developed to predict cyclone frequency, intensity, and potential impacts under different climate change scenarios. Participatory research methods, such as Community-Based Participatory Research (CBPR) and Participatory Mapping, involve local communities in the research process, fostering community ownership and enhancing the relevance and applicability of the findings. Economic and livelihood analysis is conducted through cost–benefit analysis, livelihood impact assessments, and epidemiological studies to understand the health impacts of cyclones. Health vulnerability mapping is created to identify areas with high health risks due to cyclones, integrating factors like healthcare access, population density, and environmental hazards. Policy and governance analysis is also conducted through policy impact evaluation and governance network analysis. These methodological innovations not only improve the quality and reliability of research findings but also ensure the research is grounded in the realities of the affected communities, making it more relevant and actionable.

6.4.3 Contribution to Theory

The study "Cyclone-Induced Vulnerability and Migration in Indian Sundarban" contributes significantly to various theoretical approaches across multiple disciplines, enhancing the understanding of complex interactions between natural disasters, human vulnerability, and migration. Key theoretical contributions include climate change and disaster theory, which integrates environmental, social, economic, and political dimensions specific to cyclone impacts in coastal regions. Resilience theory is introduced by identifying adaptive strategies and coping mechanisms employed by communities in the Sundarbans, while disaster risk reduction (DRR) theory provides empirical evidence on the effectiveness of various risk reduction strategies. Migration theory is expanded by detailing how sudden-onset disasters like cyclones trigger immediate and long-term displacement, introducing new concepts such as "environmental refugees" and "climate migrants" in the context of cyclones. Social-ecological systems theory advances the understanding of human–environment interactions by illustrating how communities in the Sundarbans interact with their natural environment before, during, and after cyclones. Socio-ecological resilience is studied by studying how integrated human and ecological systems respond to cyclone impacts, proposing models that depict the co-evolution of social and ecological resilience.

Political ecology investigates the role of power dynamics and socio-economic inequalities in shaping vulnerability and migration patterns, revealing how marginalized groups are disproportionately affected by cyclones and have fewer resources for recovery and adaptation. Resource distribution theory examines how the distribution of natural resources and access to services influences community resilience and migration decisions, contributing to theories on environmental justice and equitable resource management. Development theory integrates findings into sustainable development theory, stressing the importance of disaster resilience in achieving sustainable development goals (SDGs). Health and well-being theories enhance theories related to health vulnerability by linking physical and mental health impacts of cyclones to broader socio-economic conditions. Public health adaptation theory identifies effective health interventions and infrastructure improvements that mitigate cyclone impacts. Policy and governance theories explore adaptive governance by exploring flexible, inclusive, and participatory governance structures that can enhance resilience to cyclones. Policy implementation in disaster-prone areas contributes to theories on policy effectiveness and the role of institutional capacity.

6.5 Scope of the Future Research

Future research on *Cyclone-Induced Vulnerability and Migration* is crucial due to the increasing frequency and intensity of cyclones due to climate change. Potential areas for future research include climate science and environmental studies, social and demographic studies, economic impact and livelihoods, health and well-being, policy and governance, community-based studies, technological innovations, and interdisciplinary approaches. Climate science and environmental studies can focus on improving prediction models for cyclones, understanding their changing patterns, and the role of climate change in intensifying these events. Social and demographic studies can identify vulnerable populations within the Sundarbans, understanding socio-economic factors contributing to their vulnerability and how these vulnerabilities evolve over time. Migration patterns can be investigated, including forced displacement and voluntary migration. Economic impacts can be quantified, including damage to infrastructure, agriculture, fisheries, and tourism. Health impacts can be examined, including immediate injuries, mental health issues, and long-term health effects due to displacement and living in temporary shelters. Resilient public health infrastructure can be evaluated and strengthened with the use of strategies.

Policy and governance can be explored through disaster management policies, climate adaptation policies, community-based studies, technological innovations, and interdisciplinary approaches. Resilient infrastructure and early warning systems can be developed, while interdisciplinary approaches involve developing integrated models that combine environmental, social, economic, and health data. In conclusion, future research on *Cyclone-Induced Vulnerability and Migration* must adopt a holistic approach, integrating various disciplines and methodologies. By addressing the multifaceted challenges posed by cyclones, researchers can contribute to sustainable development and improved livelihoods for the communities in this unique and vulnerable region.

6.6 Concluding Remarks

The findings of this comprehensive study underscore the critical need for targeted, multi-dimensional policy interventions to address the complex interplay of climate change, migration risk, and socio-economic vulnerability in cyclone-prone villages of West Bengal. This region, particularly the Sundarban area, faces a heightened risk of cyclonic storms, embankment breaches, and soil salinization, all of which have profound impacts on agricultural productivity and local livelihoods. A significant proportion of villages in the area—66%—are exposed to moderate-to-high hazards, while 72% face a high-to-very high risk of embankment breaches. This situates the region as a focal point for policy planning aimed at mitigating risks and reducing the pressure for migration.

Agricultural Vulnerability and Food Security

The study identifies a key challenge: the region's dependence on agriculture, particularly paddy cultivation, as the backbone of food security and livelihoods. The traditional bimodal cropping pattern, aligned with the Kharif and Rabi seasons, has been compromised by increasing soil salinization and frequent inundation. This environmental shift has led to a significant reduction in Rabi crops, particularly in southern regions, weakening the area's agricultural resilience. Cyclones such as Aila and Bulbul have severely impacted Kharif crop yields, emphasizing the vulnerability of monsoon-dependent agriculture to extreme weather events. The trend of decreasing winter cropping due to salinity and increased flooding underscores the need for agricultural innovations that can adapt to changing environmental conditions. Improved irrigation systems, crop diversification, and soil management techniques should be prioritized to ensure that agriculture remains viable, thus reducing the socio-economic pressures that drive migration.

Migration Dynamics and Socio-economic Vulnerability

Migration in this region is not solely driven by environmental factors; socio-economic conditions play a critical role. The study reveals that 59.77% of the villages face high-to-very high migration risk, particularly in vulnerability hotspots such as Gobardhanpur and Krishnadaspur. Proximity to the coast, low elevation, frequent cyclone paths, and cropland vulnerability make these villages more susceptible to the impacts of natural disasters, pushing many residents toward migration as a survival strategy. Social and economic vulnerabilities exacerbate these risks, with factors such as low literacy rates, the prevalence of Kutcha (temporary or non-permanent) structures, and limited income options heightening the likelihood of migration. The study highlights that young people, particularly men with secondary education from joint families, are more likely to migrate, especially in search of work in manufacturing and construction industries. Migration appears to offer temporary economic relief, as these individuals return home periodically, particularly during cultural festivals like Durga Puja, but they remain highly exposed to the risks posed by cyclonic storms when they return to their native villages. This pattern reveals the precarious balance between the economic benefits of migration and the physical and socio-economic vulnerabilities that remain unresolved in their home regions. **Vulnerability of Marginalized Communities**: The study underscores the disproportionate impact of migration on marginalized communities, particularly SC (Scheduled Castes) and OBC-A (Other Backward Classes—Category A) families, who are more likely to migrate due to a combination of economic hardships, poor housing conditions, and limited access to social and financial capital. These households often live in mud houses, rely on loans for subsistence, and face additional challenges such as housing damage and crop or fishing losses. The study further notes that families living within 500 m of the coast or riverbank are particularly vulnerable to migration pressures due to their higher exposure to natural hazards. This highlights the need for targeted policy interventions that consider the specific vulnerabilities of marginalized and geographically isolated populations.

Migration Risk Reduction

The study's findings highlight several key policy recommendations for reducing migration risks in cyclone-prone villages. Policies should prioritize the development of climate-resilient agricultural practices, including crop diversification, improved irrigation, and soil management techniques. Support for farmers in adopting these practices through training, access to credit, and subsidies for resilient crops could enhance food security and reduce the economic pressures driving migration. Strengthening infrastructure, particularly embankments and cyclone-resistant housing, is essential for reducing vulnerability to natural disasters. The introduction of low-cost insurance schemes could provide a financial safety net for families, helping them recover from disaster-related losses and reducing the need to migrate for economic reasons.

Expanding vocational training and employment opportunities in non-agricultural sectors, such as manufacturing, construction, and service industries, could reduce reliance on agriculture and fishing, which are particularly vulnerable to cyclones. By providing alternative livelihood options within the local region, policies can help stem the tide of migration and create more stable, long-term economic opportunities for residents. Targeted interventions should focus on improving the living conditions and economic opportunities for SC and OBC-A families, who are disproportionately affected by migration pressures. This includes providing access to better housing, education, and social protection programs. Policies should also address the social and economic impacts of migration itself, ensuring that migrant workers have access to social protection, healthcare, and education, both in their destination areas and upon their return to their native villages. Supporting seasonal or temporary migration as part of a broader livelihood strategy can help mitigate the negative impacts of migration while enhancing the overall resilience of affected communities.

References

Arias P, Bellouin N, Coppola E, Jones R, Krinner G, Marotzke J et al (2021) Climate Change 2021: the physical science basis. Contribution of Working Group I to the Sixth Assessment Report of the Intergovernmental Panel on Climate Change; technical summary

Bhattacharyya A, Werz M (2012) Climate change, migration and conflict in South Asia. In: Rising tensions and policy options across the subcontinent. Center for American Progress, Heinrich Boell Foundation, Berlin

de Siqueira JH, Mtewa AG, Fabriz DC (2022) United Nations Development Programme (UNDP). International conflict and security law: a research handbook. TMC Asser Press, The Hague, pp 761–777

Gakpey EE (2022) Understanding the dynamics of climate change, migration and conflict: the case of Bangladesh

International Organization for Migration (2011) World migration report 2011: Communicating effectively about migration. IOM. https://publications.iom.int/system/files/pdf/wmr2011english.pdf

IOM (2018) Migration, environment and climate change: definitional issues. International Organisation for Migration, Geneva, Switzerland. https://www.iom.int/migration-and-climate-change

Jha CK, Gupta V, Chattopadhyay U, Amarayil Sreeraman B (2018) Migration as adaptation strategy to cope with climate change: a study of farmers' migration in rural India. Int J Clim Change Strat Manag 10(1):121–141

Kenfack CE (2022) The Paris agreement revisited: diplomatic triumphalism or denial of climate justice?. J Environ Protect 13(2):183–203

Luetz JM, Merson J (2020) Climate change and human migration as adaptation: conceptual and practical challenges and opportunities. Clim Act 120–132

Macintyre A (2020) Adaption to data-driven practices in civil society organizations: a case study of Amnesty International. J Inform Tech Polit 17(2):161–173

Mashi SA, Oghenejabor OD, Inkani AI (2019) Disaster risks and management policies and practices in Nigeria: a critical appraisal of the National Emergency Management Agency Act. Int J Disast Risk Reduct 33:253–265

UNHCR (2021) Global report 2020. United Nations High Commissioner for Refugees

World Bank (2013) The world bank annual report 2013. The World Bank

Index

P. Mandal and S. Paul, *Climatic Vulnerability and Migration: Evidence from Indian Sundarbans*, Springer Geography, https://doi.org/10.1007/978-3-032-05858-4

MIX
Papier aus verantwortungsvollen Quellen
Paper from responsible sources
FSC® C105338

If you have any concerns about our products,
you can contact us on
ProductSafety@springernature.com

In case Publisher is established outside the EU,
the EU authorized representative is:
Springer Nature Customer Service Center GmbH
Europaplatz 3, 69115 Heidelberg, Germany

Printed by Libri Plureos GmbH
in Hamburg, Germany